Apprentice to Genius

Apprentice to Genius

The Making of a Scientific Dynasty

ROBERT KANIGEL

THE JOHNS HOPKINS UNIVERSITY PRESS

Baltimore and London

Printed in the United States of America on acid-free paper

Hardcover edition published in 1986 by Macmillan Publishing Company
Johns Hopkins Paperbacks edition, 1993

The Johns Hopkins University Press
2715 North Charles Street
Baltimore, Maryland 21218-4319
The Johns Hopkins Press Ltd., London

Library of Congress Cataloging-in-Publication Data will be found at the end of this book.
A catalog record for this book is available from the British Library

For Judy and David

Contents

Introduction

HE WAS A WITTY, charming, superbly groomed man with sparkling eyes and enormous presence. He thought nothing of calling you at three or four in the morning to run his latest idea by you. His name was Bernard B. Brodie, but everybody called him Steve.

It seems that back in 1886, a twenty-three-year-old New York saloonkeeper had jumped from the Brooklyn Bridge into the East River to win a two-hundred-dollar bet—and lived to collect it. His name was Steve Brodie, and the expression "to pull a Brodie" promptly entered the language as meaning to attempt a dangerous stunt or take a long shot.

Sixty years later, during the time when Dr. Bernard Brodie presided grandly over his lab at Goldwater Memorial Hospital in New York, he had already won a reputation for taking scientific long shots. "Let's take a flier on it," he'd say. By this he meant to try an experiment that stood little chance of succeeding, but promised a big payoff if it did. In recognition of this streak in him, somebody began calling him "Steve," in memory of the bridge jumper, and the name stuck.

"Let's take a flier on it" was what Julius Axelrod heard all during his years as a technician in Steve Brodie's lab. Later,

Axelrod broke away from Brodie, got his own lab, and became famous in his own right. Then his students learned from their mentor what he had learned from his.

Many years later, a young professor of biochemistry at George Washington University Medical Center in Washington, D.C. was talking of his days as a graduate student, just recently past, and of the senior scientist under whom he'd worked, Candace Pert. "She was always willing to take a long shot," he was saying. "That's just her style."

Except it was not just *her* style. It was hers, and her mentor's before her . . . all the way back to Steve Brodie: Candace Pert was a product of Solomon Snyder's laboratory, Snyder was one of "Julie's boys," and Julius Axelrod had apprenticed under Brodie.

•

Science remains one field in which something like the traditional master-apprentice relationship still prevails. Such relationships are often intense, with both caught up in the fever of their common work, keeping long hours, sharing the triumph of a successful experiment or the frustration of one that fizzles. Through it, the student is trained in the ways of the mentor, and comes away with an approach, a style, a taste in the mouth or a feel in the gut for just what makes "good science." Through it, favors are granted, careers advanced, the sway of a particular scientific discipline extended. But through it, too, resentments sometimes spark, lifelong bitterness is kindled.

Popular wisdom fancies science a largely solitary pursuit, coolly isolated from the hot passions of human intercourse. Before 1969, when I met my first "real" scientists, I'd thought that, too. But in that year, I became friends with a graduate student in biology at Johns Hopkins University in Baltimore.

The picture my friend drew of her experiences in the lab was hardly confined to experiments, papers, and data. Her doctoral advisor was a Professor von Ehrenstein, and von Ehrenstein's lab became, in Isabelle's telling, a complex social organism with a distinct flavor and ambience. There, in that tight, circumscribed community, I heard described, and came to see myself, a whole world of friendships and rivalries and fierce, Nobel Prize–fed ambitions, of lab parties and love affairs, of mid-experiment pizza forays, of gossip traded outside the electrophoresis room, of competition for equipment and for the lab director's attention, of young love and unabashed enthusiasm and deep, festering anger.

That was my first glimpse of science as a social, and profoundly human, activity. Science was not just ideas and instruments, test tubes and lab notebooks. It was personal interaction as intense as any among actors or combat soldiers or, indeed, any group in which sustained and intimate contact stirs high feelings.

Years passed. It was 1981. I had just begun researching an article about a Hopkins neuropharmacologist, Solomon Snyder, who though barely forty had already emerged as an internationally renowned researcher. It was our first interview and it was going poorly. I asked him about his discoveries, and he told me. But my questions touched off no sparks. He was courteous and correct, but his replies remained formal, his face impassive. I was getting nowhere. And then I remembered. . . .

In preparing for an interview, it is good journalistic practice, and plain common sense, to review what else has been written about your subject in newspapers and magazines. Now, sitting across from Snyder in his comfortably furnished office at Johns Hopkins, I recalled something I'd encountered more than once, if fleetingly, in my readings the night before: Snyder, it seemed, had gotten his start as a scientist in the laboratory of Julius Axelrod, a 1970 Nobel laureate in medicine. And so, on a long shot, I steered the interview away from Snyder's scientific accomplishments. "What was it like," I asked, "working for Dr. Axelrod?"

His face exploded with delight. "Oh, it was very exciting," he sighed, tone and color in his voice for the first time. "It was wonderful." Whereupon he proceeded to recount his days with Axelrod two decades before.

The turnabout was startling. It saved the interview. But

more important, it left me with a story bigger and more ambitious by far than any I'd envisioned.

Some time before, Elise Hancock, my editor at *Johns Hopkins Magazine*, for which I'd written frequently over the preceding five years, had had an idea for an article about the role of mentoring relationships in science and academia. For a serious university magazine like *Hopkins*, which aspired to *New Yorker*-like reportage of academic and scholarly subjects, it was a natural. Hancock had even started a folder on the topic, which had just begun to receive scholarly attention.

But that was as far as the idea had gone. The problem was, how could you get at it, journalistically?

Now, in the wake of my interview with Snyder, I saw a way to give life to the abstraction that was the mentor relationship. Snyder had been so deeply influenced by this man, Axelrod, that now, almost twenty years later, the mere mention of his name stirred him to fond reminiscence. Here was the perfect flesh-and-blood match to Hancock's idea. She assigned me to the story.

I went to talk to Axelrod at his lab at the National Institutes of Health in Bethesda, Maryland. A sweet, white-haired man who, now past seventy, could have been anyone's favorite uncle, Axelrod obliged me with impressions of Snyder from their time together. But soon, with an unguarded intensity reminiscent of Snyder's, he was telling me of his own beginnings as a scientist and the role played in his life by a giant presence of a man named Bernard B. ("Steve") Brodie. Brodie had, until his retirement ten years before, been one of the most famous pharmacologists in the world.

My story was turning out to be more richly folded than I'd first imagined. Yes, Snyder had been shaped by Axelrod. But Axelrod's scientific past, it now seemed, had been enriched by a similarly towering figure. The mentor chain didn't stop with Axelrod but reached back at least one more generation, to Steve Brodie.

Later, I learned of Brodie's own roots as a scientist, and how he'd influenced not just Axelrod but a whole generation of pharmacologists, making him the linchpin of many such mentor chains whose links today reach around the world. And the same could be said of Axelrod and Snyder. They were distinguished scientists. But more, they'd "bred" other scientists who'd themselves attained renown. Indeed, I was to learn that among the scientific elite such chains of "hereditary" influence are no aberration. They are the norm.

And I learned, too, that far from cool and heady, these deeply personal relationships often burn with intensity. In pursuing my story, first for the *Johns Hopkins Magazine* article and then for this book, I experienced something so often in interviewing the scientists involved that it became almost predictable: I had but to broach the name of the person who had molded him or her as a scientist, and any cold recital of facts would cease. The voice would soften, or quicken, or rise to anger, or otherwise fill with feeling. One's mentor, I found, was rarely a neutral subject.

One time a young scientist, Gavril Pasternak, was telling me about his personal and professional debt to his mentor, Sol Snyder. "I owe everything I do to him. I try to emulate him completely," he said, adding: "Professionally, Sol is my father, and in a way I consider Axelrod my grandfather."

Did others in Snyder's lab take such intense, unblushing interest in their scientific lineage? Was that how they talked by the coffee machine at lunch, or as they waited for the data to come off the scintillation counter?

Oh, yes, Pasternak replied, "*We were always interested in our genealogy.*"

And for good reason: "Genealogy" plays as central a role in the careers of scientists as it once did in the alliance-by-marriage diplomacy of the royal houses of Europe. A scientist's early reputation rests almost as much on whose lab he has worked in—on whose scientific progeny he is—as on what he has discovered. There are "schools" of science, in particular disciplines, just as there are in art and music. There are scientific "families," each of whose members can be traced to one or a few original Adam-like figures. By one reckoning, more than half of all American Nobel prize winners have worked as graduate students, postdocs, or junior colleagues of other Nobel laureates.

The line of scientific inheritance between Julius Axelrod and Sol Snyder, running "up" to Steve Brodie, and "down" to Candace Pert, typifies mentoring networks prevalent in the highest ranks of science. Each of these men and women is firmly settled among the elite of their field. Each has made landmark contributions. Each has been singled out for numerous prizes and honors. Each has been mentioned as having done work of Nobel Prize quality, and at least three of the four have been nominated for the prize. One got it.

Each link of this generational chain served as scientific "father" to the next. Each was first a student, apprentice, protégé; each then assumed the role of mentor to the next in line. Each, by virtue of experience, senior standing, and example, guided and influenced the more junior person, passing on lessons learned, instilling in him or her a sense of how the game is played—and how it's won.

Nobel Prize winner P. B. Medawar once wrote, in *Advice to a Young Scientist*, that "any scientist of any age who wants to make important discoveries must study important problems." But what makes a problem "important"? And how do you know it when you see it? The answers don't come from reading them in a book, nor even by explicitly being taught them. More often, they're conveyed by example, through the slow accretion of mumbled asides and grumbled curses, by smiles, frowns, and exclamations over years of a close working relationship between an established scientist and his or her protégé.

This is a book about one such interlocking chain of mentor relationships.

Apprentice to Genius

1.
Nobel Laureate

THEY'D CALLED HIM from the lab that morning to tell him that Julie Axelrod had won the Nobel Prize.

But what about Steve? Costa wondered.

Erminio ("Mimo") Costa was Steve Brodie's old second-in-command at the Laboratory of Chemical Pharmacology, Brodie's great research fiefdom spread across the seventh and eighth floors of the Clinical Center, the centerpiece of the National Institutes of Health (NIH) research complex in Bethesda, Maryland. For years scientists from all over the world had flocked to his lab just to work beside him, sample his frighteningly original mind, and absorb the raw, electric energy of the place.

Now it was October 1970, Brodie was retiring, and Costa had his own lab. But the two were still close. They'd met at a scientific meeting in Miami in 1959, each of them arriving at the hotel desk to learn their room reservations had failed to go through. They commiserated, shared a cab to a motel, and talked science late into the night. Ultimately, Brodie invited him to join his lab. Costa resisted; he'd heard how overpowering Brodie could be and how demanding. Still, he came—and, as he says, "never regretted it."

Costa felt he owed Brodie a great debt of gratitude. "I consider him responsible for what I am," he would say years later. It was Brodie who had taken under his wing a virtually unknown, thirty-six-year-old pharmacologist from Cagliari, Italy, at the time stuck at an out-of-the-way lab in the midwest. Brodie had imbued him with an addiction for discovery, had taught him the place of the imagination in science.

Brodie was, by universal acclaim, the Father of Drug Metabolism—the science of how the body absorbs, transforms, and renders safe or useful the chemicals it takes in. He had broken open the field, forging it into a real science. It was out of Brodie's lab that so many dozens of distinguished pharmacologists had emerged, their careers forever altered by their exposure to his iconoclastic intellect and awesome force of personality.

And Julius Axelrod? Why, he had started out, years back, as Brodie's technician, the hands for Brodie's mind. He was a quiet, self-effacing sort of fellow who, midway through his thirties, still had no doctorate, the union card of a scientist, and who'd apparently been content to toil away in a food testing lab all his life. Until, that is, he'd come under Brodie's spell, back when Brodie was still at Goldwater Memorial Hospital. "Before he joined up with Brodie," Costa would note, "nobody knew anything about Axelrod. It was only then that he started to bloom."

Since then, there'd developed a simmering resentment between the two, culminating in the big break over credit for the discovery of microsomal enzymes. Axelrod had finally earned his Ph.D., and gone on to make major breakthroughs in neuropharmacology. Yes, Costa felt, he'd come a long way, and no doubt was worthy of the prize. *But what about Steve?*

This morning he was supposed to pick Brodie up at his apartment and drive him to the airport. Had he heard the news yet? Costa wondered.

He drove up to the ten-story apartment house on Battery Lane that had been Brodie's home for most of the past decade.

It stood within a few hundred yards of the southern edge of the NIH campus, and Costa had been here many times for late-night work sessions. It was Brodie's lab-away-from-the-lab, where he wrote his papers and developed many of his ideas. Costa parked, entered the lobby of the imposing, gray-and-white brick structure, walked down the hall to Brodie's first-floor apartment, rang the bell, and stepped in.

Brodie was sitting there, alone.

The two of them waited, silently, while Anne, Brodie's wife, got ready.

Finally, Costa asked, "Did you hear yet?"

Silence.

Based on the early news reports out of Stockholm, there'd been some question about who was to share the prize with Axelrod. The Nobel Prize in Medicine is usually divided among up to three recipients. Besides Axelrod, one was the distinguished Swedish neuropharmacologist Ulf von Euler. But the other, according to the earliest reports, was some unnamed British scientist. When Brodie did at last speak, it was to inquire who it was. Costa told him it was Sir Bernard Katz, a German-born biophysicist who'd fled the Nazis in 1935.

"He was not mad that Axelrod got it," says Costa. "It was just, 'He got it and I didn't.' "

Three years before, Brodie had won the Lasker Award, the most prestigious American award for biomedical research, carrying with it a ten-thousand-dollar check, an engraved citation, and a replica of the Winged Victory of Samothrace. Brodie was euphoric. There'd been an elegant luncheon at the Saint Regis Hotel in New York, the whole room brimming over with flowers; it was the kind of affair only Mary Lasker, wife of advertising executive Albert Lasker and force behind the award, could pull off with such style. Senators and congressmen were on hand. And the huge news conference that went with it was a press agent's dream.

On no fewer than sixteen previous occasions, Lasker Award winners had gone on to win the Nobel Prize. And the Nobel

Prize in Physiology or Medicine, to give it its proper name, was what Brodie dearly coveted. His friends and colleagues all knew as much. In 1969, when President Johnson awarded him the National Medal of Science, one of Brodie's old colleagues wrote him congratulations, adding, "Now if we can only get you that other one. . . ."

And now his former technician, meek little Julie, whom Brodie had rescued from obscurity and turned into a real scientist, was walking away with the Big One.

"It should have been Steve," someone recalls Axelrod saying after he'd heard the news. Others thought so, too—or, rather, thought it should *also* have been Steve; no one questioned Axelrod's claim to a piece of it.

•

He was in the dentist's chair, his mouth stuffed with cotton swabs, when he learned it was true. Because of his dentist's appointment, Julius Axelrod had skipped breakfast and missed the eight o'clock radio news to which he usually listened. When he arrived, his longtime dentist, Dr. Ben Williamowsky, told him he'd won the Nobel Prize.

"In what?" asked Axelrod.

"Peace," smiled Williamowsky.

"Then I *knew* he was kidding," says Axelrod.

In fact, he didn't wholly dismiss the possibility of a Nobel Prize. Like most scientists, "I'd dreamed about it. I thought I had a remote chance." When, a little later, while already in the chair with his mouth full of cotton, the nurse came in and said that a radio station reporter wanted to know how he felt about winning the Nobel Prize, he realized that Williamowsky hadn't been kidding after all.

"I was flustered, excited, my heart was pounding," he remembers. It was October 15, 1970.

He drove the seven or eight miles from his dentist's Silver Spring office to his lab at NIH in Bethesda. He looked for a place to park. He couldn't find one. Axelrod, who is forever being described as a "sweet" and "good" man, was just this once bad. "To hell with it," he remembers thinking, bypassing the clogged parking lots and pulling right up to the entrance of the Clinical Center, the twelve-story, multi-corridored labyrinth of laboratories, offices, and patient care facilities in which he'd worked for fifteen years.

By this time, the corridor outside 2D45, Axelrod's lab, had filled with dozens of his research associates, friends, and colleagues. Calls were coming in from everywhere. When Axelrod appeared—tieless, in a baggy, checked short-sleeved shirt, dark pants, and suede loafers, looking a little the worse for his visit to the dentist's—he was, as one report had it, "roundly cheered." Someone had sent out for a bottle of champagne. Now it was opened, and everyone stood around, smiling and happy, drinking out of paper cups.

The new Nobel laureate was soon being whisked off to a noon press conference at the Barlow Building, the agency's headquarters. There, before cameras and clumps of microphones, he fielded the inevitable questions about the significance of his work and what it might someday mean for treatment of the mentally ill. When one reporter asked him to spell *norepinephrine*, the sympathetic nervous system neurotransmitter whose workings he had helped reveal over a period of fifteen years, he was too rattled to get it right.

Later, President Nixon called him, telling him that he had been a source of "great pride to all of your fellow citizens," and an example of the "outstanding efforts to improve the physical and mental health of mankind to which the United States is dedicated."

"It was like talking to a phonograph," Axelrod remembers. Still, he used the opportunity to appeal for relief from proposed cuts in the federal budget for basic research.

Like a proud parent, Axelrod's employer since 1955, the National Institute of Mental Health (NIMH)—which is administratively separate from, but functions closely with, the National Institutes of Health—was understandably pleased at the honor bestowed on one of its own. And for November 3, two weeks after announcement of the Prize, it and NIH planned a joint recognition ceremony for Axelrod. Announcements were sent out, guests invited, flowers ordered, videotaping and sound taping arranged, special parking arrangements made, even a seating chart prepared for on-stage dignitaries.

Among the seven invited to appear on stage with Axelrod was Steve Brodie.

John Eberhart, NIMH's director of intramural (in-house) research, wrote Brodie: "Because of the important part that you personally and the Heart Institute generally have played in the research area for which the prize was given, and because of Dr. Axelrod's earlier association with your laboratory, we would like to invite you to be present. . . .

"I do hope you can accept. I am sure it would please Dr. Axelrod."

A few days later, Brodie called. Yes, he said, he'd be there.

•

"It's satisfying to see one of our boys win the Big One," said John Eberhart at the ceremony, which jammed to capacity the Clinical Center's Masur Auditorium.

The prize "couldn't have happened to a nicer guy," said one old colleague of both Brodie's and Axelrod's from their New York days. Someone else described Axelrod as "a most generous colleague, a superb and generous teacher."

Was it hard for Brodie, sitting up there and listening to his scientific competitor and former student acclaimed for having won science's highest honor? "I'm sure," says a former colleague, "that deep down in his soul he was thinking, 'It should have been me. Why didn't they put me up there?' "

"One of Steve's frustrations was that he wasn't recognized" with the Nobel Prize, says John Burns, Mimo Costa's predecessor as Brodie's deputy. "He had a tendency to take things personally, and that put him into a depression."

Today, Brodie will only say, a big, toothy grin erupting across his face, that he was "surprised" at the Nobel committee's announcement.

Was he envious?

"That's a hard question," he replies, still smiling broadly. "I'd say I was surprised. Many people were surprised."

Like Herbert Weissbach, who worked in Brodie's lab along with Axelrod and now is director of the Roche Institute of Molecular Biology. Weissbach feels Brodie should have gotten the prize right along with Axelrod. "I think it would have been perfectly fitting," he says. "Then no one could have said, 'How come?' "

Or like Elliot Vesell, another graduate of Brodie's lab (and Brodie's distant cousin), who is now a professor of pharmacology at Hershey Medical Center in Pennsylvania. To him, the case of Brodie recalls that of Oswald T. Avery, who in the 1940s showed that DNA is the genetic material and set the stage for the discovery of DNA's double-helical structure by James Watson and Francis Crick—but who never received the Nobel Prize for his crucial work. Vesell sees "political" factors at work: "Brodie is so large and dynamic in the field. He dominated it. . . . People may have been turned off by his aggressiveness. He did have enemies."

During the ceremony, Brodie's name came up several times. At one point he was introduced as "a scientist and teacher of scientists whose contributions have been so many." But he stood only briefly in acknowledgment, said nothing, then sat promptly down, as the rain of accolades for his former technician continued. "Brodie didn't look very happy," says one of those there that day. "His facial expression seemed depressed. The others on stage were ebullient."

•

In the second row of the auditorium, just behind Axelrod's wife Sally, sat a young former student of Axelrod's, Solomon Snyder.

Just as in any family tree many branches connect back to the trunk, so can many lines of mentoring influence be traced back to Brodie. But of them, none are stronger than that reaching down, through Axelrod, to Sol Snyder. And none have left so deep a mark on science. Axelrod is Brodie's most famous scientific progeny; Snyder is Axelrod's. Snyder met Brodie face to face just once or twice. Yet some spark that Brodie passed to Axelrod—dimmed in some ways, strengthened in others, but in the end largely intact—was handed down to Snyder. And he in turn would bequeath it to his own students.

Snyder was the classic *wunderkind*, brilliant and fiercely ambitious. At the almost unseemly age of thirty-one, he was already a full professor of pharmacology and psychiatry at Johns Hopkins. Ahead was his codiscovery of the opiate receptor—direct proof that the brain contains molecules specifically tailored to recognize opiates like heroin—which would bring the glare of TV lights and reporters from *Time* and *Newsweek* to his lab at Hopkins. He would win literally dozens of top scientific awards, including the Lasker. He would author or coauthor hundreds of scientific papers, as well as popular books; would be named head of a department at Hopkins reputedly created just for him; would be mentioned persistently as a candidate for the Nobel Prize.

Snyder loved Axelrod. "He'd always been the closest to him of all of Axelrod's students," says one of Snyder's own former students. "He worshipped him and lionized him, and did for him." It had been Axelrod who had deflected him from a career as a conventional psychiatrist and hurled him into the heady world of research. Research wasn't something gray and drab, Julie had taught him; it could be fun, exhilarating, an adventure into the unkown. And for that lesson he was indebted to him. The day Julie won the prize, Snyder's ex-student recalls, he was "beyond ecstasy." He sent Axelrod this telegram:

> THRILLED BEYOND WORDS TO HEAR ABOUT YOUR NOBEL PRIZE. MAZEL TOV 10,000 TIMES. BEING YOUR STUDENT WAS SO WONDERFUL AN EXPERIENCE FOR MYSELF AND FOR OTHERS YOU SHOULD GET A SECOND NOBEL PRIZE AS A TEACHER OF SCIENTISTS.
>
> SOL

The preliminaries were over. Axelrod was introduced. At first, as the standing ovation washed over the auditorium, he stood uneasily at his chair, fussing with his papers. Then, in short, shuffling steps, he walked to the podium.

Ever since the prize announcement two weeks before, it had been like this. Attention. Adulation. Winning the prize, he would say, was "like being made a cardinal." Reporters pouncing on him for interviews. Telegrams: "Heartfelt congratulations on this great honor," read a typical one. "All of us who know you and your superb work share in the satisfaction of this moment. Never has an honor been more fully earned and well deserved." Among those sending congratulations was one Achilles M. Tuchtan, mayor of Rockville, Maryland, the new laureate's hometown. Another was the president of Axelrod's high school alumni association, who told of a huge sign being hung in the auditorium to honor this Seward Park High School alumnus, class of 1929.

Ahead was the reception at the Swedish Embassy in Washington. The appeal from a New York congressman for Axelrod's views on "the most effective formulas for achieving a lasting world peace." The hundreds of letters from sick people and their loved ones from around the country desperately seeking medical advice. (And the letters back explaining that the winner of the Nobel Prize in Medicine was not, in fact, a medical doctor.) And then, in Stockholm, for the Nobel festivities, aided by written instructions on how to comport himself: lunch with the American ambassador, the great banquet for seven hundred at city hall, the traditional Festival of Saint Lucia (the Swedish Queen of Light), the pretty girl with the crown of candles in her hair serving him breakfast in bed, the gold medal presentation from the king of Sweden, and on and on.

Now, from the podium in Masur Auditorium, just downstairs from the laboratory in 2D45 where he'd made most of his discoveries, he could see many of his friends and colleagues from over the years at NIH. Even Brodie was here today.

"My fantasy," he'd confide many years later, "was winning it with Brodie."

It was almost a quarter-century ago that they'd started together. Back in World War II, the government had mounted a crash program to develop antimalarial drugs, and Brodie, working with a group of young researchers at Goldwater Memorial Hospital in New York City, had been a key figure in it. After the war, on the advice of his boss, Axelrod had approached Brodie for help on a problem. A couple of weeks' collaboration endured for nine years, and what they found together broke open the science of pharmacology.

But always it had been Brodie, the senior researcher, the lab director, the brains, the guiding force, while he, Axelrod, was the technician. He was a gifted one, certainly, a bustling vortex of energy—but always, in Brodie's eyes, just a technician.

Until finally he had grown resentful, even bitter, in the wake of the microsomal enzymes discovery, and gone his own way. For nine years, they'd worked at each other's sides virtually every day; now, in the fifteen years since the break, they'd seen each other, for even so much as a hello, maybe a dozen times. And this was one of them.

"First I want to thank Steve Brodie," he began, going on to tell the story of how they met. "I went to spend an afternoon and stayed nine years." Laughs from the crowd. "After that I couldn't see any other career but research."

•

His talk that day was one extended expression of appreciation. And among those he thanked were "all my former colleagues at Goldwater Memorial Hospital who gave me my first taste of what a creative research environment was like."

It was Goldwater where Axelrod had met Brodie and gotten his start. But in a real sense, American preeminence in biomedical research could trace its roots there as well.

Three decades separated Axelrod's Nobel Prize in 1970 from Steve Brodie's crucial, war crisis–fueled discoveries in the basement of Goldwater in 1941. In 1941, the United States as a scientific power was still second to the great research centers of Europe. In 1970, it was, by any yardstick, first.

In 1941, science was still largely the province of a few gentlemen investigators. In 1970, it was a profession all its own, attractive to the best minds, open to everybody.

In 1941, the National Institutes of Health awarded twelve research grants worth, in all, seventy-eight thousand dollars. In 1970, it awarded 11,339 grants valued at more than six hundred million dollars. Whereas thirty years before, a "government job" as a scientist was open to derision, by 1970 a position in the NIH intramural research program was widely coveted. By then, NIH was already established, as the distinguished molecular biologist Donald Brown puts it, as "the most cost-efficient, best-run, most effective agency of government. No other organization gives as much to the world."

The man universally credited with the rise of NIH was a tall, bespectacled Irishman with a propensity for speaking in mumbles and mutters named James A. Shannon. Shannon directed NIH from 1955 to 1967, but was already making his presence felt there by 1950. "One of the most remarkable leaders of science in this country," a former colleague has called him. "He turned around scientific research in this country. He made NIH into NIH."

But before Shannon made NIH into NIH, he made Goldwater into Goldwater. It was he who directed the antimalarial program there. He was Steve Brodie's boss; he *hired* Brodie. He became Julius Axelrod's boss. It was he who fashioned what Axelrod called, in his Masur Auditorium comments, Goldwater's "creative research environment."

So that in the mentor chain whose first links were forged in the basement of Goldwater can be seen not only the workings of elite science generally, but also the whole growth and maturation of the American biomedical research enterprise since World War II. During one brief span, a few years at most, Brodie, Axelrod, and Snyder all worked just an elevator's ride away; appropriately, they were all part of the Shannon-directed NIH of the mid-1960s, a time and place of explosive growth and buoyant optimism in American science.

"There's a reason that this is the mecca of medical science," observed Lewis Aronow, chairman of pharmacology at the Uniformed Services University of the Health Sciences, gesturing around him at the array of health research complexes on both sides of Rockville Pike in Bethesda. "There's a reason that English is the language of science in the world today. It's Jim Shannon and what he started at Goldwater."

2.
A Wartime Urgency

IN MARCH 1942, four months after the bombing of Pearl Harbor and American entry into World War II, the Japanese seized the Dutch East Indies, cutting off virtually the entire world's supply of quinine.

Quinine, made from the bark of the cinchona tree, was the standard treatment for malaria, the mosquito-borne disease that killed three million people a year worldwide and incapacitated hundreds of millions more.

To American troops in the South Pacific, malaria loomed as dangerous as the Japanese. In places like New Guinea, rains fall almost nightly, filling every hollow with puddles in which breed the *Anopheles* mosquito, whose bite transmits the malarial parasite. The GIs had mosquito nets under which to sleep while in camp, small head nets while in the field. They were issued repellents and mosquito bombs. Yet still they came down with the shaking chill, the terrible fever—sometimes as high as 106 degrees—and the sweating, headache, nausea, and appetite loss. Typically, bursts of intense fever would be followed by seeming remission, only to recur a week or two later.

Malaria had weakened more than half the American soldiers

surrendering at Bataan early in the war. In later campaigns, whole army divisions were rendered unfit to fight. If the Japanese were to be beaten, it became plain, the various species of *Plasmodium*, the parasite that *Anopheles* passes to its human host, had to be overcome first. Malaria would be termed "the number one medical problem of the war." But even before Pearl Harbor, in the spring of 1941, it was being accorded special attention by the National Research Council, and by mid-1942 a crash program for the development of antimalarial drugs had been launched.

It was a massive program, coordinated by a loose network of panels, boards, and conferences. The actual work was conducted at dozens of universities, hospitals, industrial laboratories, and army and navy facilities. By war's end, some fifteen thousand potential antimalarial compounds had been screened and tested. Extracts from Chinese herbs received serious consideration. So did mud from the River Nile, juice from the leaves of the cotton plant.

At Johns Hopkins University in Baltimore, canaries, chicks, and sixty thousand ducks were infected with malaria, their responses to various drugs closely monitored. At the Illinois State Penitentiary at Joliet, convicts watched as physicians let mosquitos alight on their bellies and bite them. Federal and state prisoners in Atlanta, New York, and New Jersey also served as human guinea pigs. So did conscientious objectors and victims of central nervous system syphillis—for whom malarial fever was, as it happens, a long-accepted treatment.

The focal point for the clinical testing of drugs emerging from the various screening programs was the Research Service of the Third Medical Division of Goldwater Memorial Hospital in New York City, under the direction of James Shannon.

•

People who worked for him would later say, virtually without exception, that Jim Shannon was the best boss they'd ever had. "I don't know anybody who worked for him who didn't idolize him," says Jack Orloff, who occupies one of Shannon's former posts, as director of intramural research at the National Heart, Lung, and Blood Institute, one of eleven institutes today comprising the National Institutes of Health. One day Shannon would have NIH's central administration building named for him. He'd have awards and testimonial dinners showered on him, and more than two dozen honorary degrees. Most everybody liked Jim Shannon, at least those not intimidated by his diamond clear intellect, or put off by his cool, austere demeanor. He would be held up as a model scientific administrator. He would be called "a medical and scientific phenomenon."

But at the time he took over at Goldwater in 1941, Jim Shannon was just a young assistant professor of physiology at New York University. He had no administrative experience to speak of, and his area of scientific expertise was not malaria but renal physiology, the study of the kidney.

Shannon, J. A. The excretion of inulin by the dog. *American Journal of Physiology*. 112: 405–413, 1935.

Shannon, J. A. Glomerular filtration and urea excretion in relation to urine flow in the dog. *American Journal of Physiology*. 117: 206–225, 1936.

Shannon, J. A. Urea excretion in the normal dog during forced diuresis. . . .

Open up a physiology textbook even today and Shannon's contributions, half a century later, still get prominent mention. "Almost singlehandedly, he was able to transform renal physiology from an observational qualitative science into a highly precise quantitative one," is how one appreciative account of his career summed up his accomplishments.

The kidneys, each the size of a fist in humans, cleanse the blood of waste products and otherwise regulate its composition. But it's by no means clear from looking at their microscopic tubules and their meshwork of filtering units and tiny blood vessels, called glomeruli, just how they work. Back when Shannon was still a bench scientist, it was even less clear.

Through a series of thirty-five papers, Shannon helped clarify how the kidneys form urine. He developed ways to monitor the effect on them of various drugs and hormones. He developed the use of inulin, a starchlike substance made from dahlia tubers, as a probe of renal function—a seemingly narrow methodological finding that, in fact, profoundly affected future work in the field.

By the late 1930s, Shannon had "established himself," in the words of Thomas Kennedy, a longtime friend and colleague who by the 1980s was director of policy and planning for the Association of American Medical Colleges in Washington, D.C., "as one of the brightest young [scientific] guys in the United States. He was a real golden boy."

A golden boy who was, as Kennedy says, "getting restive." In 1934, Shannon had married. Now he had two young children, born in 1937 and 1939, and even in the Depression, thirty-six hundred dollars a year didn't go far. Perhaps, Shannon let the word get out, a department chairmanship was opening up somewhere? None were.

Then John Wyckoff, dean of the NYU School of Medicine, came along with an offer. It was thanks to Wyckoff that Jim Shannon had got into medical school in the first place. Back in college, at Holy Cross, Shannon had distinguished himself mostly on the basketball court, as the team's star center, and on the track, as a cross-country runner. He was "a goof-off in college," according to Tom Kennedy, and Shannon remembers worrying, just ten days before he was supposed to graduate, whether he was going to be kicked out of school for some minor infraction. He "barely got into medical school," says Kennedy. The interview with Wyckoff was what did it. Wyckoff, it seems, had had a similarly checkered undergraduate career, took a special liking to the young man from Holy Cross, and squired him to a place in the class of '29.

Now it was 1941, and Wyckoff had another opportunity for his young protégé. Would Shannon be interested in heading up NYU's Goldwater research service? He'd have, Wyckoff assured him, all the resources he could want, including lots of both patients and lab space. And the salary would be ten thousand dollars a year. Shannon needed no convincing.

•

"It is inadvisable," noted a report to the New York City Commissioner of Hospitals in 1934, "to treat large numbers of the chronic sick in any one general hospital . . . [because invariably they] suffer and are exploited for the benefit of those acutely ill." Needed, therefore, was a hospital expressly for the chronically ill. And in July 1939, the first such hospital in the country opened its doors on the site of an old penitentiary on a long, narrow island in the middle of the East River between midtown Manhattan and Queens.

Set out in a series of four "pavilions" arranged like sergeant's chevrons and linked by a single central corridor, Goldwater Memorial Hospital extended some eleven hundred feet along the southern end of Welfare Island, almost exactly in line with East Fifty-seventh Street in Manhattan. This design, according to an early history of the place, gave patients "a maximum amount of sunlight and fresh air and afford[ed] for each one a view of the busy and interesting channels of the East River."

Also notable was that, by law, this new sixteen-hundred-bed hospital was to include a number of research divisions, each to be linked to one of New York's medical schools, and each to have beds specifically allocated to it. One would be tied to Columbia's College of Physicians and Surgeons, another to Cornell (though Cornell never actually participated). The third research service was assigned to New York University, and on January 1, 1941, James Shannon became its director.

Eighteen months later, Shannon's career in renal physiology would be over and he'd be a novice student of malaria.

•

The first thing Shannon did was get busy recruiting. For forty years, Shannon's eye for scientific talent has been legendary: The roster of young Ph.D.s, physicians, and even lab technicians whose distinguished careers he helped launch is long and impressive. Goldwater was a preview of what he would do, on a larger scale, at the National Institutes of Health.

One of those he brought to Goldwater went on to win a Lasker Award for his work in a field he almost singlehandedly opened up. Another became deputy director for science at the National Institutes of Health, and from there dean of a top medical school. A third helped found, and built into a respected institution, a private biomedical research institute. Of Shannon's fifteen or so key people at Goldwater, a veteran of those days has reckoned, all but one or two wound up as research directors, university department chairmen, and the like.

Goldwater was so new and fresh its research labs in Building D weren't even finished yet; Shannon was starting with a clean slate, and that helped his recruiting. So, too, did his own scientific reputation. Finally, he already knew a lot of people, friends in high places able to funnel him likely prospects.

None of which explains the magic Shannon brought to identifying promising researchers in the first place. Certainly, though, he had it—had, as one admirer once put it, a "wholly inexplicable capacity to pick the right people." Had, in the words of future Nobel laureate Christian Anfinsen, who was among those Shannon later recruited to NIH, "a golden touch."

But in 1941, Shannon's young team was still mostly unproved. They had just barely set to work expanding on his renal physiology work from the 1930s when the attack on Pearl Harbor propelled America into World War II. Wartime medical priorities would hardly include kidney physiology near the top. "Rather than breaking up the group, getting commissions [in the army or navy], and treating casualties," Shannon figured, "we'd do better to put our scientific talents into war research."

What kind of war research was settled when Shannon, on a trip down to Washington, D.C., stopped in Baltimore to see his old Johns Hopkins friend, pharmacology professor E. K. Marshall.

Ken Marshall and Jim Shannon went way back. Marshall, one of the country's leading pharmacologists, had for years had a place up in Maine, where he'd worked summers at the Mount Desert Island Biological Laboratory, along with Shannon and Shannon's old lab chief at NYU, Homer Smith. Smith, whom Marshall had "discovered" back in World War I, was a famous kidney physiologist whose lab Shannon had joined in 1931, after completing his Bellevue residency. Shannon remained there nine years, picking up a Ph.D. to go with his M.D., and making many of the contributions to renal physiology for which he became known. "The brightest star in the constellation of Homer Smith," someone would later call him.

For six productive summers, Shannon worked with Smith at the Mount Desert Island lab. There he met, and became close friends with, Marshall. Aloof and stiffly formal, Marshall was "difficult to know, a cold man in interpersonal relations," says Marshall's biographer, pharmacologist Thomas H. Maren, also a Mount Desert Island veteran. But he had a soft spot for Shannon. Marshall's wife once told him, Maren reports, that Marshall had "more affection for Shannon than anyone he's ever known."

Now, in 1941, Shannon was visiting him in Baltimore. They spent the evening together, at Marshall's big, three-story gabled house in a woodsy northern area of the city. "Why don't you," Shannon remembers Marshall asking in his Charleston drawl, "come into the malaria program with me?"

Marshall, fifty-two at the time, was a prime mover in the big government program. He was a consultant to the national drug screening panel, ran his own busy lab in Baltimore, and served on the Board for the Coordination of Malarial Studies, based forty miles down the road in Washington, D.C. Now he wanted Shannon to join him.

Shannon protested: Why the urgency? Didn't we, despite the Japanese seizure of the world's quinine plantations, have ample stocks of Atabrine, a synthetic antimalarial developed by the Germans during the 1930s?

Yes, Marshall replied, there was plenty of Atabrine to go around, but the troops wouldn't take it. It made them sick. Turned their skin yellow, too. And it didn't work all that well in the first place. Why, they were pulling troops off Guadalcanal, site of a big early offensive against the Japanese, because of Atabrine toxicity.

All right, Shannon said, he'd put his group to work on the problem right away.

•

"Cure for Malaria is Revealed after 4-Year, $7,000,000 Research," the *New York Times* declared in a front-page article on April 12, 1946. "The curtain of secrecy behind which the multimillion dollar Government antimalaria program had been operating, in the most concentrated attack in history against this scourge, was completely lifted today for the first time, with the revelation of the most potent chemicals so far found."

Later events proved talk of an outright cure for malaria premature; malarial strains have developed resistance to each new drug developed. But to troops in the field, it was as good as a cure. In an editorial the next day, the *Times* said that "when the scientific story of the war is written, we have here an epic that rivals that of the atomic bomb, the proximity fuse, and radar."

Most of the excitement in 1946 concerned several new drugs, developed late in the war, shown to hold great promise. But back in the dark days after Pearl Harbor, more crucial to the war effort by far was work done at Goldwater in 1942 and 1943 that retrieved Atabrine from the brink of the pharmacological scrap heap.

Atabrine had been developed by the Germans in 1932 after a long bout of research going back to World War I. American chemists synthesized it a few years later. Even before Pearl Harbor, drug companies were reportedly churning out 500 million pills a year. In 1942, science writer Paul de Kruif, first of the science popularizers and author of the bestseller

Microbe Hunters, was touting Atabrine, in the pages of *Reader's Digest*, as "the new malaria giant-killer."

Only trouble was, it wasn't true. So great were the problems with Atabrine that the army for a time halted production and prepared to discontinue its use. Many soldiers wouldn't take the drug. It yellowed the skin. Its intestinal side effects were sometimes as debilitating as the malaria it was supposed to suppress. It was even rumored among the GIs to impair sexual vigor, a belief the Japanese exploited through air-dropped leaflets. (On New Guinea, American propagandists countered with billboards showing a jolly sultan leering at a dancing girl as he popped a pill, remarking, "Atabrine keeps me going!") Worst of all, Atabrine simply failed to suppress malarial attacks, and was painfully slow to stop them when they did come. But this was how it worked at the approved dose of one tenth of a gram three times a day—a figure set with the idea of simply mimicking quinine, whose use against malaria went back at least three hundred years.

This method of setting the dose was, by today's standards, unspeakably crude and only by the most charitable use of the word "scientific." But it was about all anyone did back then: Pharmacologists would give patients a drug of given dosage and see how it affected disease symptoms. Experience with enough patients in hand might allow them to start divining what a good dose was. "Such an approach to the general problem of Atabrine therapy," Shannon and his Goldwater colleagues would note in a 1944 journal article, "is a striking contrast to the more quantitative one which has facilitated the development of sound antibacterial therapy with the sulfanilamides."

The sulfanilamide drugs, beginning with the clinical introduction in 1935 of Prontosil, were the first of the "wonder drugs" used to combat bacterial infections. In the sulfanilamide field, Ken Marshall was a key figure. He had helped develop two new sulfa drugs, sulfapyridine and sulfaguanadine. More important, he'd pioneered, and was a prime proponent of, a new, more quantitative approach to dosage-setting.

The cornerstone of this new approach was to first find a way to measure the concentration of the drug in the blood—or, to be more precise, in the blood plasma, which is the clear, colorless liquid left behind when red and white blood cells are removed. Why the blood plasma? Because it's next best to what you're *really* after, which might be the liver, the central nervous system, the kidneys, or whatever: The level of drug in the blood is close to what the particular tissue "sees."

With a means of measuring blood levels in hand, the next step was to determine what blood level achieves the therapeutic end desired. Then working backwards, and with knowledge of how the drug is absorbed, metabolized, and excreted by the body, a dosage schedule to maintain that blood level could be designed. This was the strategy Marshall had found so successful in the case of the sulfa drugs—a strategy that, as the National Heart Institute's Jack Orloff says, "everybody today thinks is obvious." Yet most pharmacologists of the day had not yet grasped this linchpin of what was to be called the New Pharmacology. To Shannon, it was already gospel.

There was only one hitch: You had to be able to measure blood levels, and that wasn't always easy. Ken Marshall had done it with sulfanilamide. But each new drug meant starting all over again, as if on a new problem. And in mid-1942, with the army about ready to give up on it, no one knew how to measure Atabrine or, for that matter, other compounds that might be formulated to replace it.

Measuring Atabrine levels in the blood, then, was the chief obstacle to success of the whole program. Two people were assigned the problem. One was a technician named Sidney Udenfriend, a twenty-four-year-old from Brooklyn just finishing up his master's degree at New York University (NYU). The other was an English-born thirty-five-year-old organic chemist whom Shannon had brought with him from NYU: His name was Bernard B. Brodie, but everyone called him Steve.

•

Brodie, Udenfriend, and the rest of the malaria group were stationed in the basement of Goldwater's Building D. It had thirty-five hundred or so square feet of space, divided into five or six small labs. There was a library, a chemical storage room, two or three small offices. As modern laboratories go, there wasn't much to it.

But the air down there fairly crackled with electricity. "There was very high esprit," recalls Tom Kennedy, who came there as an army medical officer in 1944. "Shannon was a superb leader. Everybody was first rate. Everybody was interested in research. Everybody knew this was an extremely important project." More than one veteran of those days would call it the most exciting period of their lives. "The scientific enthusiasm we shared at Goldwater," wrote one of them, George Downing, on the occasion of a reunion years later, "has continued to fuel our lives ever since."

It was a setting in which answers were needed "tomorrow," as Shannon says, where research was necessarily directed "two or three steps away from current knowledge." It was a fiercely results-hungry place, with no room for the kind of plodding, workmanlike approach that in other circumstances might have been all right. Nor was there room for sensibilities in need of isolation and quiet; the crisis demanded constant exchange of information and ideas. There was no standing on ceremony of the kind exemplified by the traditional German laboratory system, where the *geheimrat*, or "Right Honorable Laboratory Director," was apt to be saluted in the morning before he issued the day's working orders. There was a war going on. And that fact, coupled with some of the best young scientific minds of the day bent on finding answers, filled the air with excitement.

And it was Shannon who made the whole thing work. Shannon's role had changed. He was no longer the lone scientist, patiently toiling away in the lab with a couple of assistants. By the time the war was over he'd have five senior researchers, close to a dozen technicians and army and navy medical officers at his disposal. He had access to a ninety-bed ward staffed by some twenty nurses and ward attendants. For Shannon, the bench scientist, it was almost as if he'd set out on a new career. It was one in which he would make an even more distinguished mark, as a research administrator.

"A great 'bureaucrat,'" someone would call Shannon years later, putting the word in quotes as if to acknowledge that a word so mired in negative connotations had no business being applied to someone like him. But bureaucrat, of a rare sort, he had become. In his memoir *What Happened in Between*, William Welch, a physician who joined the malaria program in 1944, saw Shannon "waking to a new and not-unsympathetic capacity in himself for wheeling and dealing." He was forever rushing off to Washington, "his briefcase stuffed with data and his head with new proposals for the waiting committees to weigh and discuss in the heat of the national emergency."

Later, when already director of the National Institutes of Health for some years, Shannon told a reporter that when he first came to NIH he'd hoped to return to research one day. "I never did get back," he said, "and I suspect I won't." But in fact, the transformation had begun much earlier, at Goldwater. Sid Udenfriend remembers Shannon trying to anesthetize a writhing rat one day. The rat scratched him. "Hell," growled Shannon, throwing down the animal. "I can't do research part-time." It was, says Udenfriend with a dash of pride at having watched history in the making, Jim Shannon's last experiment.

A passport photo of him taken in 1936, when Shannon, thirty-two, was a guest investigator at the Physiological Laboratory at Cambridge University in England, shows a long, almost horsey face, dark, wavy hair, casually knotted tie, and soft eyes behind round-lensed, wire-rimmed glasses that give him a scholarly, even sensitive, air. It is the last photograph of him that looks anything like that. In almost every subsequent picture, the necktie has given way to an austere bow tie, and the soft boyish eyes have grown more steely and intense.

He was a handsome man, well built, and tall enough, at six-foot-two, for the Holy Cross basketball team. Outwardly taciturn and formal—to some, even a little cold—he was, as Kennedy says, "a superb leader," and by all accounts a natural one, the kind of man accustomed to working with and through others to achieve his ends. Even back in college, when the rest of the basketball team elected him their captain, the school newspaper had predicted that "Big Jim should prove to be a fine leader," noting his "cool generalship" when the regular captain was out of action.

Shannon was impressive in appearance, quiet, firm, strong. And, in the words of Robert Berliner, who was with Shannon at Goldwater virtually from the beginning, "smart as hell," which, in a community of bright people, counted for a lot. His mind functioned with awesome clarity, his scientific articles and reviews proceeding as if in response to an almost inevitable logic. Steve Brodie once called him "the most logical man I have ever ever known."

Unlike many leaders, Shannon's authority did not flow from a forceful speaking style. In fact, he tended to mumble, the ends of his sentences dropping off, as if into a void. "I had to know what he was saying in advance," recalls a Goldwater technician, "to know what he was saying at all."

Still, he inspired immense respect. "You could feel his presence. You knew he was the guy to be the boss. You could feel the power emanating from him." That's how the wife of one Shannon lieutenant remembers him. He had a way of getting the most out of people. One time Brodie was in Shannon's office, lamenting a problem that seemed to resist solution, when the phone rang. It was an army officer inquiring about progress on that very problem. "Don't worry," Shannon assured him, "Brodie is here. It's under control."

"He had absolute confidence I could do it," Brodie remembers with wonder. "And that *made* me do it. It was great psychology."

Shannon, as another colleague remembers of later experiences working under him, was a man whose "own ego was satisfied if the people he picked did well. . . . He got the credit eventually, of course. But he got by giving." When the war ended and it was time to publish the results of their work, recalls Tom Kennedy, Shannon made sure they all got their due, went to the important scientific meetings, got a chance to be seen by the leading names in the field.

He'd do anything to support his people. One time, Kennedy was assigned to Green Haven, an army prison whose inmates had volunteered to be injected with malarial blood. ("We asked them to volunteer, but I guess there was some implied reward," admits Kennedy.) The Green Haven hospital was supposed to furnish needed nursing help and ward attendants. "But the commanding officer was a son of a bitch. He made life hard for us. He wouldn't cooperate."

Kennedy complained to Shannon. Next thing he knew, Shannon was on the phone to a high army official and ordering a staff car to take them to Green Haven.

"He marched into that guy's office, coattails flying," recalls Kennedy, his eyes glowing as he tells the story. "He put his feet up on the CO's desk, and said, 'I was just talking to General Walton,' or whatever his name was, 'and he says. . . .' He created the impression that if he didn't shape up and fly right there'd be hell to pay. Well, after that, things went smoother."

•

It was remarkable how smoothly things went—or at least how quickly. By the spring of 1943, as General MacArthur was getting set to "leapfrog" his troops across the South Pacific against the Japanese, Brodie and Udenfriend had unlocked the secret of Atabrine and made it into a potent malaria fighter.

It is not the mosquito that causes malaria, recall, but rather the parasite with which it infects its host. The parasite—any of several species of *Plasmodium*—multiplies in the red blood cells of its host, sometimes reaching concentrations of five hundred thousand per cubic millimeter. When the cells rupture, breakdown products surge through the system, and the body responds with fever. In the case of one particular species, *falciparum*, the parasites can clog the blood vessels of the brain, causing death.

To be effective, an antimalarial agent must disrupt the parasite's life cycle. To do that, it must reach the parasite. And that means getting into the blood. Finding a method to determine how much is in the blood was the crucial problem Steve Brodie and Sid Udenfriend faced.

Now, measuring *pure* Atabrine posed no particular problem: When irradiated by light of the right wavelength, Atabrine, like many other organic, or carbon-containing, compounds, will fluoresce; that is, excited by the incident light, it will give off light of its own. This emitted light lies in the ultraviolet range of the electromagnetic spectrum, just outside the human eye's sensitivity. But though invisible to the eye, it can be readily measured with a standard laboratory instrument called a photofluorometer. And—this is the important point—the intensity of fluorescence is proportional to the concentration of the compound. So if you set the photofluorometer's zero point with a blank pool of water, and calibrate the instrument with a known sample, you can measure drug concentration directly.

But what if you don't have a pure specimen of Atabrine? Separating Atabrine from the rest of the blood plasma was, to be sure, a problem soluble by more or less standard chemical means. Distinctly less routine was distinguishing it from its own metabolites: Typically, an ingested drug doesn't stay in its original form. Some or all of it may be metabolized, or chemically changed by the body into a different form. How would you know you were measuring Atabrine and not a chemically similar cousin?

The crucial insight came when Brodie noted that metabolites of chemically alkaline drugs like Atabrine were almost invariably more polar than the original drug. (A polar molecule is electrically imbalanced; it has a heavier "weight" of electrical charge out at one or another end of it.) Couldn't that difference, he wondered, be exploited?

Polar substances dissolve well in other polar substances;

their electrical imbalance is mutually relieved, as it were, by the intimacy of contact that being in solution implies. Polar molecules, for example, dissolve readily in water, which is highly polar. Less polar molecules, on the other hand, tend not to dissolve in water, but do tend to mix well with other nonpolar liquids, like the organic solvents ethylene dichloride or benzene.

Since Atabrine was less polar than its metabolites, Brodie reasoned, maybe it could be separated from them by its lesser tendency to dissolve in water. Maybe it could be carried off by some less polar liquid while its metabolites were left behind in aqueous solution.

Much work went into refining the basic technique, such as selecting the right solvents, settling on ideal temperature and acidity for the various chemical steps of the extraction, and working out various bugs. But this was the basic strategy Brodie and Udenfriend succeeded in making work. And with the drug thus neatly separated it became relatively easy to measure, by photofluorometry, just how much of it there was.

"The Estimation of Atabrine in Biological Fluids and Tissues" appeared in the *Journal of Biological Chemistry* in 1943, but Brodie and Udenfriend had worked out the method by the winter of 1942. It could legitimately be said, as Brodie claimed years later, that their method for measuring Atabrine had "saved the day." For it meant that Shannon's group was no longer in the dark. It could now, easily and routinely, trace the fate of any Atabine given the patient. What was more, the new technique worked not just in the blood plasma, but in urine, feces, or any body tissue. And not just in human patients, but in experimental animals.

In fact, it was an experiment with a dog that yielded perhaps the most startling and significant finding of all. The dog was given ten milligrams of Atabrine intravenously. Four hours later it was killed and the Atabrine in the plasma and in various tissues was measured. Concentration in the dog's muscle fiber turned out to be two hundred times that in the plasma, and in its liver, two thousand times.

Body tissues, it seemed, soaked up Atabrine; very little of it was getting into the blood where it could do some good. If you gave Atabrine to humans at the doses then officially called for, it mostly accumulated in tissues like muscles and the liver; only slowly would it build up in the blood to the thirty-microgram-per-liter level found necessary to kill off the *Plasmodium vivax* parasites. No wonder, then, that at the dosages then being prescribed Atabrine took so painfully long to work.

The solution was clear. Doubling or tripling the regular dose might kill the parasites all right, but it would only exacerbate, to an intolerable degree, the side effects the soldiers were already experiencing. But what if, the first day of chemotherapy, you gave a big "loading dose" of the drug and, from then on, gave relatively small daily doses to maintain the blood levels thus achieved. That way, you'd saturate the tissues right away and all succeeding drug would go directly into the blood stream.

That's what they did and that's what happened. By the spring of 1943, the problem was licked, and the new dosage schedule instituted. By January 1944, as Brodie later wrote, "malaria as a tactical or strategic problem had practically disappeared."

•

"One of the most exciting periods of my career was working with you at Goldwater Memorial Hospital on the malaria project," Brodie once wrote Shannon. "That was the real beginning of my career."

But it was the beginning of something more, too. Brodie and the other Goldwater veterans had grown addicted to something in the air down in the Building D basement—a ferment, a sense of urgency and excitement. To them, for the rest of their careers, less exhilarating science would hardly seem like science at all. And that standard of what research could be at its best they'd bequeath to all who'd work with and for them. Something had taken root at Goldwater, in the fertile soil Jim Shannon had prepared and nourished, something that would branch out and grow lustily, shooting out tendrils that would root years later and miles away.

Brodie had come into the war with a respectable list of scientific credits. He came out of it a scientific star. Energized by the wartime urgency at Goldwater, his spirit brimming with confidence and his head with ideas, he was ready to pounce upon the world with a New Pharmacology largely of his own making.

3.
Steve Brodie, Methyl Orange, and the New Pharmacology

SOMETIMES, IT SEEMED, Steve Brodie never slept. For days on end, he'd get by on two or three hours a night. In the middle of a long work stretch, he'd sometimes doze off for twenty minutes, then return to work on all cylinders.

He was a nocturnal creature, at Goldwater as well as later when he was chief of the Laboratory of Chemical Pharmacology at the National Heart Institute. Typically, he wouldn't get into the lab till noon. He might go home at six or so, but then only to start a second workday after dinner—late evening sessions at his apartment writing papers, making phone calls to his colleagues at two or three in the morning, inquiring about experimental data, batting out a volley of new ideas. "He'd call anytime he felt like it," remembers Lewis Aronow, who worked with him in the early 1950s. " 'How can you say that?' he'd demand to know about the paper you'd written. 'Why don't you do it this way?' "

For Brodie, there was no separation between work and play, the lab and home; time did not break down neatly into days and hours. One time, recalls Elliott Vesell, who worked in Brodie's lab for four years during the early 1960s, Brodie was attending a scientific meeting at the Mediterranean beach resort of Saint-Tropez. He'd been lost in discussion for hours when, abruptly, he looked up, looked around, and observed quizzically, "You know, for a resort, there aren't many people around." It was three o'clock in the morning.

"Working with you was not always easy," a former colleague, Alfred Pletscher, reminded him years later, referring to Brodie's indifference to the normal working day. Once, the story goes, Pletscher was working day and night on an experiment when Brodie, impatient for results, prodded him: "*We want to get something into print.*" Replied Pletscher, looking Brodie straight in the eye: "I'm not working hard enough?"

"He was a slave driver," says Lewis Aronow.

"Psychologically demanding," says Parkhurst Shore, a longtime associate, "but he was such an exciting bloke that in a way you didn't mind it."

"I couldn't sleep when I had a good idea," is how Brodie himself explains his midnight drivenness. And when it came to ideas, he was a bottomless reservoir of them. Not all were good, but good or not they'd gush up from his imagination like Old Faithful, and with as little respect for the limits of the normal workday.

He could be immensely stimulating. "Let's pretend we are not biologists, but chemists," he would say, embarking on a round of Socratic questioning. Or, "If we were nature how would we build this [biochemical mechanism]?" Large stretches of the night might pass in this kind of reasoning. And out of it would come, almost invariably, ideas for new experiments, provocative possibilities revealed in what had seemed run-of-the-mill problems. To most—not all—who worked with him it was ample repayment for their lab chief's relentless demands, the bleery-eyed night sessions, the mind-battering intellectual combat.

It was exhausting to oppose him in argument, yet he encouraged it. His colleagues were his sounding board, and often he'd champion scientific ideas he himself didn't fully believe, the better to spark a spirited discussion. Nor were the debates always gentlemanly. Tempers sometimes flared. As one who dueled with him remembers, the discussion could be grueling, "abrading to the sensibilities."

"Steve was probably as singlemindedly focused on science as anyone I've ever met," recalls Tom Kennedy. "He lived it, breathed it, night and day." He had no hobbies worthy of the name, few interests outside the lab. Once, while he was at the heart institute, the employee newsletter profiled him. "Science," it was said of him, "is his relaxation as well as his work, relieved only by occasional entertaining, reading detective stories (the more miserable the plot the better), and the movies (indiscriminate). The interviewer could find not one homey hobby to endear our scientist to the masses."

In the years they worked together, he and Park Shore had a running debate on the value of The Balanced Life, Brodie insisting that it condemned one to mediocrity. "I always maintained that the most important thing in life is living," says Shore. "For him the most important thing in life is work."

"Brodie worked as hard and as long as anyone I've ever known," says Aronow.

"He didn't look upon it as work," says Sidney Udenfriend.

Brodie's peculiar nocturnal habits, together with his other quirks of scientific and personal style, would come into fullest bloom later, at the heart institute. But already, at Goldwater, during and after the war, there were the long nights doing experiments or writing umpteen drafts of his papers. One who worked with him closely during this period says he relied on amphetamines to help keep awake, barbiturates to get him to sleep. Whatever the source of his energy, it was awesome. "Everyone grumbled that he worked twenty-three hours out of twenty-four," recalls Betty Berger, a technician there. "And he expected us all to have the same drive, to be as excited as he was."

Brodie's secretary at Goldwater, Shirley Udenfriend, says she today prefers to portray Brodie as the classically absent-minded professor, charmingly idiosyncratic. At the time, though, she was not always so understanding. He could be insufferably demanding. His correspondence was forever piling up on his desk and it was always the day before a holiday, or so it seemed, that he'd get to it. Then, at four-thirty in the afternoon, just as she was winding up for the day and girding for the long commute back to Brooklyn, he'd show up at her desk with a pile of stuff for her to do. It was maddening.

The man was driven.

•

"They were all like that," says Betty Berger of the Brodie clan: strong, dominating personalities that roiled with intensity.

Before coming to Goldwater, Berger had gone to Barnard College in New York with Brodie's sister, Rachel. "A female Brodie," is what she calls Rachel, who became a social worker, teacher, guidance counselor, piano teacher, and mother. "She and Steve both had a devil-may-care attitude, a certain breeziness. They were really gung-ho." Brodie's brother, Henry, whom she also knew, was a career diplomat with the state department; tall, thin, and serious-looking, he reminded some of Abraham Lincoln. "The Brodies were all the same way. They were on top of the world. They were going to accomplish things. It was wonderful being with people like that."

(Of all the Brodies, some said brother Maurice was the most brilliant. He'd developed a polio vaccine back in the 1930s, and was briefly hailed for it. But when, apparently without his approval, the vaccine was actually tried on a few children, one of them contracted polio and died. Not long after, under clouded circumstances, he died, at the age of thirty-seven. Betty Berger remembers the Brodies treating the episode as "a big family secret." For his part, Steve Brodie says his brother died of a heart attack, while playing golf.)

At the time of her friend Rachel's graduation from Barnard, Berger met Brodie's mother, Esther Ginsberg Brodie, who came down from the family home in Ottawa, Canada. "She gave the impression of being very firm," Berger remembers. Son Steve calls her "a very wonderful woman. We got whatever brains any of us had from her." Brodie's wife, Anne, who knew Mrs. Brodie well before she died, describes her as "a driver. 'Don't come home unless you win,' she'd say."

Early in life, Bernard Beryl Brodie showed few signs of living up to that dictum. He was born in Liverpool, England, probably in 1907—some confusion surrounds the exact year—the third of five children. When he was four, the Brodies moved to Canada, settling in Ottawa. His father, Samuel Brodie, owned a men's furnishings store. ("He was about the worst poker player I've ever seen," is about all Brodie, who was a good one, has to say about him.) As a boy, he neither excelled in school generally nor took special interest in the sciences. His high school chemistry teacher, he says, held little hope for his prospects. At one point, Brodie asked him for a summer job recommendation. The teacher refused.

Early in the last year of the regular five-year high school program, he got into an argument with the principal. He wanted to drop a course. The principal said no. Young Brodie insisted. The principal kicked him out of school.

In 1926, the eighteen-year-old dropout enlisted in the Royal Canadian Signal Corps, looking to the army to straighten him out; he apparently had much that needed straightening. For one, he was almost pathologically shy, sometimes even crossing the street to avoid a face-to-face encounter. The others in his unit would pick on him, even beat him up. One day, the sergeant major pulled him over for a little fatherly advice. "You have to fight back," he told him.

So Brodie learned to fight. More, he learned to box, became good at it, began entering bouts. Early in his first fight, he was knocked out cold; he doesn't remember anything before the referee's ". . . six . . . seven . . . eight. . . ." Finally, he lifted his rubbery five-foot-eleven frame off the canvas, only to be knocked down three times more.

He lost that fight, but in his thirty or forty others, he says, he never lost again, ultimately becoming Canadian Army champion in his weight class. He credits himself with being quick and stylish as a fighter. But he claims he never enjoyed it. "My ambition," he smiles, "was not to get hurt."

That's one army story Brodie likes to tell. The other is about his poker prowess. Seems he took out library books on poker, and on statistics, and became a formidable card player, during his three-year army stint amassing some five thousand dollars in winnings. Once, he found himself in a game with a crowd of reputed gangsters. At one point winning handsomely, he rose to leave. "Can't quit now," he was told. He sat down, played some more, managed to hold on to most of what he'd won, and finally, after a decent interval, got up again to leave. Not until he was out of the room, the door shut firmly behind him, did he stop shaking.

Bankrolled largely by poker winnings, Brodie enrolled at McGill, Montreal's fine English-language university. There, he gravitated toward the sciences but still showed no signs of academic distinction. Then one day in his fourth year, his life changed.

Back as a freshman, Brodie had fallen asleep during a chemistry lecture and been kicked out of class. Now, three years later, the same chemistry professor, W. H. Hatcher, stopped him on a snowblown Montreal street corner, said he needed help with an experiment, and asked whether he'd be interested. Sure, said Brodie.

Hatcher's experiment required twenty-four-hour-a-day monitoring, and the professor's wife, Brodie later learned, had had it with his late-night absences. "I was the only sucker he could find," Brodie smiles. Night after night he stayed up recording data. He was fascinated by the experiment. But more, he was intrigued by how a scientist gets ideas for such experiments in the first place. His mind found its true home. His marks improved, C's becoming A's. Meanwhile, his contribution earned him a place on the final paper:

> Hatcher, W. H. and Brodie, B. B. Polymerization of acetaldehyde. *Canadian Journal of Research*. 4: 574–581, 1931.

It was the first of more than four hundred.

Brodie applied for fellowships at graduate schools in the United States. Four sent acceptances, apparently impressed by his research experience. He entered New York University in 1931, earned a Ph.D. in organic chemistry four years later, and went to work as a research assistant in pharmacology in the lab of George B. Wallace.

•

One day in 1940, Eugene Berger, a twenty-one-year-old fresh out of Lafayette College with an undergraduate degree in chemistry, walked into the dean's office at the NYU medical school and walked out with a job in George Wallace's lab. Berger knew little about what he wanted to do in life except that, one, he didn't want to go into his father's flour and feed business back in Hazleton, Pennsylvania, and two, he needed a job. Brodie's technician had quit that day, and someone was needed to wash lab glassware. "If I'd come the day after, or the day before, it would all have been different," says Berger, who is Betty Berger's husband. As it is, he would come to share a Manhattan apartment with Brodie and, after the war, work at Goldwater, where he'd remain until 1974.

Berger counts the period in Wallace's lab as the greatest in his life. Brodie taught him how to pipette—to transfer measured quantities of solution from one test tube to another. He taught him how to use a balance. They worked out ways of assaying, or systematically measuring, calcium and magnesium in the body, looked into how those minerals affect the sleep cycle of dogs. "To me," Berger recalls, "Brodie was a god. I thought he was the most wonderful person in the world." But he was also a tough taskmaster. "You did exactly what he said. There was no room for deviation."

Wallace's lab was located in the College of Medicine, an 1897-vintage brick structure at First Avenue and East Twenty-sixth Street in Manhattan. To get there, you'd take a rickety elevator up to the top floor and come out into a huge dissecting room, usually filled with cadavers, through which you had to pass to reach Wallace's lab. Wallace, Berger remembers, delighted in seeing his visitors white-faced.

A tall man in his early sixties, trained in Europe, Wallace

was one of the country's leading pharmacologists. Among those in the department was Otto Loewi, a Nobel laureate whom Wallace had helped spirit out of Nazi Germany. For Berger, the high point of each day was the conversation over lunch with Brodie, Wallace, Loewi, and the others. "I don't remember a word that was said or an incident that occurred except that it was wonderful."

Wallace and Brodie got along well, and did solid work on how various halogens—the class of chemicals comprising fluorine, iodine, bromine, and chlorine—are distributed through the body, which at the time was an important area of research. Wallace taught Brodie about the role of intuition in science and stressed the creative freedom granted by a tentative, working hypothesis: You don't have to be sure you're right about an idea. You don't even have to be pretty sure. Rather, it's enough to go in with a good hunch, see what experiment it suggests, then test it out.

Brodie, an organic chemist by training, credited Wallace with making him into a pharmacologist, with taking in a stranger to the field, as it were, someone almost bereft of biological knowledge. Today pharmacology is pervaded by the structural formulas and molecular manipulations of organic chemistry. But in those days the two were quite distinct fields. Wallace was one of the few who could see that they'd one day have to thoroughly merge.

The lab was on the sixth floor of the medical school. Down on the fifth, until he left to go to Goldwater, was Jim Shannon's lab. From what Shannon had seen and heard of him, Brodie was not just a good chemist but someone with a mind of idiosyncratic cast not chained to the scientific conventions of the time. In 1941, he offered him a position at his new lab, and Brodie took it.

The two of them clicked. For one thing, Shannon made little effort to constrain his crazy hours, as Wallace had. For another, as Brodie says appreciatively, he "could draw me out, like Socrates, solving a problem right on the spot." Brodie respected Shannon enormously, by all accounts viewing him almost like a father. Tom Kennedy pictures the two of them as "mutually reinforcing." In embracing Shannon's gospel of blood levels—the central importance of being able to measure drug levels in the blood—Brodie was at first little more than an extension of Shannon. But his conversion was so natural and enthusiastic that soon he'd outstripped Shannon himself in devising new quantitative methods.

In any event, by mid-1942 Shannon's group had shifted its attention to antimalarials and Steve Brodie was trying to come up with a way to measure Atabrine.

•

One day around this time Brodie ran into Gene Berger, his old technician, on the street. Following the outbreak of the war, Berger had left Wallace's lab, gone to medical school, and now was well along in his studies. He and Brodie, it turned out, were both looking for a place to live. They decided to look together.

Brodie found them a place at the Beaux Arts Apartments. "It was a fascinating place to live. It was *very* interesting," says Berger, in the cryptic shorthand men sometimes use to suggest certain early adult experiences. The Beaux Arts Apartments were a pair of new, sixteen-story brick apartment buildings set across East Forty-fourth Street from one another between First and Second avenues, near the site now occupied by the United Nations. Ostensibly intended for artists, they were, of course, out of reach for most real working artists. Instead, they were populated by models, photographers, and prostitutes. And by Brodie and Berger.

Their furnished, third-floor apartment in the north building was really one large room, with a tiny kitchenette and two beds built into the wall. Their arrangement was that the first to retire for the night would pull down the bed for the other. One night, Brodie was out on a date and Berger figured there was no need to pull down the second bed. "Well," he recalls, "Steve didn't make out the way he thought he would." Around three in the morning, Berger heard him fumble with his keys, stumble into the apartment, and, a little inebriated, flop into bed . . . except that there *was* no bed, so he crashed to the floor instead.

That was not typical, however, of Brodie's luck with women. More often they were crazy about him, remembers Berger. He was charming and handsome, with dark, hooded, penetrating eyes. A friend from even before the Goldwater days, Joseph Post, describes him as resembling George Gershwin, "only better looking." One summer he and Brodie took a memorable auto trip up into Canada. They saw Brodie's mother in Ottawa, then drove through Quebec, staying in a little French town upriver from Quebec City, finally returning to New York via Cape Cod. Brodie was a wonderful traveling companion, Post recalls, with a great sense of humor, who met people easily—especially women.

John Burns, who joined Brodie after the war, and who now is vice president for research at Hoffmann-LaRoche Pharmaceuticals in Nutley, New Jersey, also remembers Brodie as infinitely enchanting to women. One time Burns introduced him to his first wife. She came away, says Burns, "just fascinated."

•

Still, women were at best second in Brodie's affections during these years. First was work. Day and night he worked.

By mid-1943, with the new dosage regimen for Atabrine established, Brodie and the rest of the Goldwater group were busy evaluating new antimalarial drugs: Though Atabrine worked at keeping soldiers fit for combat, it was no cure-all. It prevented *falciparum*-induced malaria, snuffing out any infection already present. And it suppressed attacks of *vivax* malaria; but it did not ward them off in the first place, and it could not cure infections once they had taken root.

The search for new drugs made Goldwater a larger operation. The government pumped in money, as much as Shannon wanted. Army medical officers were his for the asking. And the new routine ease of taking drug levels, made possible by

Brodie's pioneering work, helped put the project on almost a mass production basis. They'd gather in the syphilitics for whom malarial fever served as treatment, innoculate them, count their parasites, start a drug, record its effectiveness, and compare it to its predecessors.

The vast body of accumulating data was recorded on standard summary sheets listing the results by drug and species of malarial parasite. A patient named Bayley, for example, was recorded as suffering a *vivax*-induced fever for five days. Before treatment, he showed parasite counts of 22,400 per cubic millimeter. Then he got an initial dose of .075 gram of a compound called SN 7618—it was the 7618th drug tested in the national program—followed by a regular dose of .025 gram every twelve hours for five days. Its blood plasma concentration, as determined by methods devised by Brodie and Udenfriend, was 10 micrograms per liter after the first day's treatment, rising to 16 by the third—enough to register a complete cure.

A white crystal known as chloroquine, SN 7618 was one of the new drugs on which the *New York Times* lavished its front-page treatment in 1946. Chloroquine achieved results similar to Atabrine, but did so faster, needed to be administered less frequently, and caused fewer side effects. The other success story, pamaquine, not only suppressed, but permanently cured *vivax* malaria. Both drugs came along too near VJ Day to have an impact on the war, but were used for many years thereafter.

With the war over, so was the emergency that had brought the malaria group together. One by one, young medical officers, like Tom Kennedy, went back to finish their residencies. Budding researchers, like John Baer and Sidney Udenfriend, drifted back to graduate school. "The war had interrupted my career," remembers Udenfriend. "Now I was going back. It never even entered my mind not to." The day the first atom bomb was dropped on Hiroshima, in August 1945, Udenfriend applied for admission to the NYU doctoral program.

Shannon, meanwhile, left Goldwater in 1946 to become director of the Squibb Institute for Medical Research, part of the giant pharmaceuticals company. He, his wife, and their two school-aged kids moved across the Hudson to Metuchen, New Jersey, where they lived in a white shingled house in the shade of a big linden tree.

But not everyone left Goldwater. Some stayed, carrying on work begun during the war or going off in new directions spurred by the malaria project. Among them was Steve Brodie.

•

Seventeen months after the end of the war, in January 1947, Brodie and Udenfriend submitted for publication six papers collectively entitled, "The Estimation of Basic Organic Compounds in Biological Material." At first glance, this collection of "methods papers," as they were called, merely detailed certain lab techniques useful for the analysis of drugs and other organic compounds, and seemed to amount to little more than a bunch of recipes. But their importance extended far beyond their superficially narrow scope. Occupying forty-five pages of a single issue of the *Journal of Biological Chemistry*, they represented a set of powerful biochemical tools for probing the body's response to drugs.

The origins of one of those tools went back to almost the first days of the malaria program. Brodie tells the story this way:

Before Atabrine had been rescued from disfavor, the government weighed the use of bark from certain South American cinchona trees. This bark consists of four alkaloids (a class of organic, nitrogen-containing compounds found in plants) together known as totaquine. One of the four alkaloids was quinine, but at concentrations far below that found in the cinchona trees now controlled by the Japanese. What about the other three? Might they pack an antimalarial punch also? To find out, a means of measuring them in blood plasma was needed, and Brodie was assigned the job.

One of the three alkaloids yielded easily, but two—cinchonine and cinchonidine—resisted. There was nothing about them, it seemed, to measure. Gas chromatography wasn't around then. Neither of the two would fluoresce. What to do?

Brodie camped out at the big central library on Fifth Avenue, back across the East River in Manhattan. For three or four days he was there, almost continuously, reading, poring through the literature. Finally, he came upon the vast body of German dye research. Could he, he wondered, *dye* the compound, then use its intensity of color in solution as a measure of its concentration?

The compounds he hoped to measure all lay in that large segment of the chemical world known as *basic*. In chemistry, the word doesn't mean "fundamental," or anything like it, but simply places a substance on the opposite end of the spectrum from an acid: An acidic solution is high in free positive charge, a basic one in free negative charge.

When an acid combines with a basic compound, or base, it forms a salt; common table salt is a product of just such a chemical union. Similarly, when an acid dye combines with a basic compound like those Brodie sought to measure, he conjectured, perhaps it would form a dye-carrying salt.

Brodie called pharmaceutical and chemical supply houses, requesting any acid dyes available. The war emergency speeded cooperation and within a few days he and Udenfriend had hundreds. They set to work: One after another, they tried coupling them with cinchonidine and cinchonine. In two or three instances, the resulting salts did show color. But the color soon faded. That wouldn't do.

Then one night they ran out of dyes. It was two o'clock in the morning. They had been working for thirty-two hours straight. "We've failed," Brodie thought. They were all set to go home when a fierce thunderstorm struck. They *couldn't* go home. There they were, stuck waiting out the storm, stuck with their failure, when Brodie's gaze fell on the shelf beside him. There stood a bottle of methyl orange, a common laboratory reagent used to record, by a precipitous change in color, a solution's change in acidity. It was so familiar in the laboratory that they'd never thought of using it. Maybe methyl orange would form salts with those obstinate alkaloids. It was worth a try.

Five minutes later and totaquine was theirs! One had only to extract the drug from plasma by techniques already worked out, shake it up with methyl orange to form a salt, then place a sample of the resulting solution in a colorimeter, a standard lab instrument that measures the amount of light a solution transmits. Appropriately calibrated, the colorimeter then directly read the concentration of the drug.

For Brodie, the methyl orange technique was a turning point. In 1947, "Estimation by Salt Formation with Methyl Orange" was one of the six methods papers in the *Journal of Biological Chemistry* that would help establish his reputation. But more important, the experience had convinced him that, with imagination and hard work, he could measure *anything*.

Mere cookbook work, bereft of fundamental significance? Brodie didn't look at it that way. True, methyl orange, or any other technique for that matter, by itself cast no light on natural processes. Still, once you could measure a drug, you could trace its fate in the body, learn how fast it was metabolized, or how much of it was excreted, or how much collected in which body tissues, or anything else you wanted to know about it. The methyl orange technique furnished one tool for measuring drugs. Other methods yielded other tools. *Many* methods—like those represented by the 1947 methods papers —and you had the makings of a pharmacological revolution.

And it *was* a revolution. Lewis Aronow, who, with Avram Goldstein and Sumner M. Kalman, is the author of *Principles of Drug Action*, a pharmacology textbook, explains that before Brodie, a drug's potency was typically measured in terms of physiological variables, like its influence on blood pressure, say, or muscle strength. Earlier still, you gave a drug and noted whether the patient vomited, or sweated, or urinated, or bled—or died.

Brodie and his legion of followers studied drugs as *chemicals*—specifically as chemicals that both worked *on*, and were worked on *by*, the body. The lab he set up at NIH a few years later would be called the Laboratory of Chemical Pharmacology, and the choice of name was not an accidental one.

This chemical approach made more feasible than ever before the synthesis of new drugs. Until recently, most all drugs derived from natural sources. Typically, folklore told of a leaf, say, or a root with unusual properties that, ground up and appropriately treated, yielded a partially purified extract. Atropine, for example, the powerful parasympathetic nervous system antagonist, is derived from the leaves and roots of the belladonna plant. The heart stimulant digitalis comes from the dried leaf of the foxglove plant.

Long before Brodie, organic chemistry had progressed to a point that made the synthesis of new drugs possible. But before Brodie, there was so little *basis* for synthesizing them. How, with as pitiful a knowledge of drug metabolism as then existed, could you synthesize a new drug and expect it to work any better than the old?

Over the next decade, Brodie and his coworkers changed all that. By suitable chemical methods, for which one or another of the 1947 methods papers often served as a starting point, they were able to track the metabolic fate of a wide variety of drugs and, in several instances, develop better ones.

One such success story involved procaine, first synthesized in 1905 and introduced under the trade name of Novocain, the familiar local anesthetic used by dentists until only recently. By the time procaine came to Brodie's attention, however, it had been used for much else besides. For one, it had an antiallergic effect. For another, it had been used on victims of cardiac arrhythmia, a condition marked by wild, erratic beating of the heart that often leads to death. But there was a serious problem with procaine when used on cardiac patients: It stopped working too soon.

Brodie and two coworkers decided to trace procaine's metabolic fate. Using modifications of methods he and Udenfriend had worked out previously, they were able to show that the body broke down procaine rapidly, eighty percent of it being transformed into its metabolic byproducts within two minutes. One of those metabolites, diethylaminoethanol, also produced an anti-arrhythmic effect, and for a time it became the focus of their attention. But it just wasn't potent enough. Using diethylaminoethanol by itself, they found, required a dose so large it caused an unacceptable drop in blood pressure.

Was there some way procaine could, in effect, be made "stronger"—more resistant to metabolic breakdown? Brodie looked to its ester bond. An ester is a chemical group, containing carbon and oxygen in a particular configuration, which tends to break down in water. As it happens, enzymes in the blood plasma hastened that breakdown, thus cutting the molecule in two and rendering it ineffective. Now, there seemed no evidence that the ester region of the molecule played any role in the drug's short-lived pharmacological action. So, they speculated, what if the ester bond were replaced with some other, hardier kind of chemical link?

Brodie enlisted the aid of a drug company, Squibb, which synthesized a number of "variations on a theme" of procaine. The one that finally worked looked just like procaine except that where once there had been the ester's oxygen atom now there were nitrogen and hydrogen. Organic chemists call such a configuration an amide, and procainamide was what the new compound was christened. It wasn't very different from procaine, just different enough to make a difference. Whereas procaine was rendered ineffective in minutes, ninety-five percent of the procainamide originally administered was still in the plasma, still exerting its anti-arrhythmic effect, nineteen hours later.

Today procainamide is available, and widely used, under a variety of trade names. On at least one of the Apollo missions it was, as a precautionary measure for the astronauts, taken to the moon. Brodie has been quoted as saying, in reference to the American Heart Association, which supported some of the procainamide research: "They certainly got their money's worth on that one."

•

The first of the procaine papers appeared in the *Journal of Pharmacology and Experimental Therapeutics* in 1948. Earlier that year in the same journal appeared another Brodie paper that, in the long run, was perhaps even more significant. For it was the first product of a scientific collaboration that was to leave its mark on pharmacology for the next decade. It was called "The Fate of Acetanilide in Man," and it reported on the research of Bernard B. Brodie and Julius Axelrod.

4.
Brodie and Axelrod: "Let's Take a Flier on It"

ONE DAY IN 1946, Julius Axelrod's boss came to him with a new and difficult problem. The Institute for the Study of Analgesic and Sedative Drugs wanted to know why certain headache remedies, including Bromo-Seltzer (as then constituted), sometimes caused headache, dizziness, diarrhea, and anemia among their users. It was a problem more challenging by far than any the thirty-three-year-old chemist had yet tackled. Would he care to try it? Before Axelrod could respond, his lab chief put in, "I can get you help."

A tall, pipe-smoking man of seventy-two, Axelrod's boss—actually, he was the lab's president, an honorary position—had, until his recent retirement, been chairman of the pharmacology department at New York University. He was George B. Wallace, and the "help" he had in mind for Axelrod was Steve Brodie.

Axelrod called Brodie and the two of them met at Goldwater Memorial Hospital on Lincoln's Birthday, 1946. "That was a fateful day for me," Axelrod remembers. For three hours they talked. When he left, Axelrod's old life was over and his new one had begun.

At the time, Julie Axelrod was a scientific nobody. He'd grown up on Manhattan's Lower East Side, the son of Isadore and Molly Axelrod, 1906 immigrants from Galicia, an area now largely within the Soviet Union but at the time part of the Austro-Hungarian Empire. His father was a basketmaker who sold to grocers and flower merchants. Once, when he was thirteen, young Julie tried making baskets. He hated it. More fun were Saturday expeditions with his father, when the two of them would set out together in the horse and wagon, calling on customers, and sometimes Julie would get a chance to drive. Isadore Axelrod worked hard all his life. In time, he even had his own shop. But the business was always a marginal one, and whenever he did find himself with a little extra money, he'd gamble it away—a source of endless frustration to Julie's mother.

Axelrod spent the first twenty-four years of his life in a cold-water flat at 415 East Houston Street. He remembers taking baths in a wash tub in the kitchen. "But I never felt deprived or anything," he says. The immediate neighborhood was almost wholly Jewish, and he grew up speaking Yiddish and attending cheder, the after-school religious exercises which he loathed, and from which he often played hookey. A few blocks away were Ukrainian and Polish neighborhoods and once, when he wandered into one of them, some kids grabbed his hat and beat him up. "Sheeny bastard," they called him.

He attended P.S. 22, a school dating back to the Civil War. Then it was off to Seward Park High School, over on Broome Street. Seward can claim its share of illustrious graduates, including entertainers Zero Mostel, Walter Matthau, and Tony Curtis. But academically, the atmosphere was nothing like that at Stuyvesant or Townsend Harris, elite public high schools to which Julie didn't even apply. "That's where all the smart kids went," he says.

More central to Julie's life than school was the library a block from home. When he was seven, he got his first library card, and soon was headed there every afternoon after school.

He read voraciously. Sometimes he'd sit at mealtime, eating, listening to the little crystal radio he'd built, and reading a book, all at once. He read *everything*. As a boy it was *Pinocchio*, which he devoured again and again, and the Grimm fairy tales. Later came Tolstoy, Dostoevsky, and H. L. Mencken. Then, at college during the early thirties, when he flirted with various left-wing causes—"We thought the revolution was just around the corner," he says—it was Dos Passos, and John Reed, chronicler of the Russian Revolution, and many others.

In those days, City College was, as Axelrod describes it, "the proletarian Harvard," a tuition-free intellectual haven for thousands of bright immigrant kids who couldn't afford to go anywhere else. There, he majored in biology and chemistry. Back in high school, he'd thrilled to romanticized accounts of medical research, like Sinclair Lewis's fictional *Arrowsmith* and Paul de Kruif's *Microbe Hunters*. But to actually *become* a scientist? Why, as a career, there was hardly such a thing as "science" back then. Rather, studying science was just a part, and a minor one at that, of becoming a doctor. And *that* was what Julie's mother wanted to see him become. She was supportive of his studies, jubilant when he came home from school with good grades, urged him on to medical school.

He applied to several, got turned down by all. In those days, medicine was a gentleman's profession, the preserve of the well-off. Most medical schools, as studies later confirmed, had quotas on the number of Jews they would admit; Axelrod feels sure this was why he was rejected. "You had to be outstanding if you were a Jew. I was good, but not outstanding." He earned no A's among his science courses at City, only B's and C's, and even a pair of D's in math.

Axelrod graduated in 1933, into a job world plunged into depression. Factories were laying off workers everywhere. Unemployment was on its way up to twenty-five percent. Axelrod took the test for a post office position—at forty dollars a week, quite a plum—and got it. And then didn't take it.

Instead, on the twenty-five dollars a month he got as a volunteer at the Harriman Research Laboratories of New York University, he managed to survive long enough to work into a regular job as a lab assistant. When Harriman closed in 1935, Axelrod landed a position in a nonprofit laboratory being set up by the city to test food and vitamins, sort of a local version of the federal Food and Drug Administration. It was called the Laboratory of Industrial Hygiene, and Axelrod worked there for ten years.

His title was "chemist": He was more than just a technician, but then again, he did nothing of what today would be called research. Axelrod's job was to modify existing methods for the testing of vitamins, like D, A, B_1, and B_2; the idea was to standardize them, make them reproducible. "Vitamins were a big thing back then," he recalls; the city was trying to protect consumers against fraudulent and misleading claims.

One of his projects was to refine a test for vitamin D, which protects against rickets, a bone disease in which cartilage fails to calcify: A rat would be fed a diet designed to give it rickets, then administered vitamin D. Later, when the rat was killed, a thin slice of one of its bones would be dipped in silver nitrate that, when exposed to light, would reveal a dark line, corresponding to the onset of calcification induced by the vitamin. The intensity of the line could be used as a visual measure of vitamin D. The procedure itself was old hat; even such a detail as, say, the exposure time to the silver nitrate had been worked out long before. Axelrod was not being paid to be original.

Still, he had no complaints; most of his former classmates were stuck in downright menial jobs, when they worked at all. Back at the Harriman, one hot summer day in 1934, a bottle of ammonia had exploded in his face, blinding him temporarily in both eyes and permanently in the left. So when the war came, his 4F draft deferment kept him out of it. He stayed at the lab for the duration, by now with a master's degree in chemistry he'd earned at night at NYU.

It was a good, solid job. There were interesting scientific journals around, which he read with relish. The pay was all right. The work was modestly challenging. Of course, it was all he'd ever known.

•

Lincoln's Birthday 1946, and Axelrod was still at the Laboratory of Industrial Hygiene. But just now he had this tough new problem from the analgesics institute to grapple with and, at the urging of his boss, had gone over to Goldwater to talk with Dr. Brodie about it. "What's the active principle in these headache powders?" Brodie asked.

Axelrod replied that it was acetanilide, a white crystalline substance first introduced as an analgesic and fever-reducer back in 1886. "Put its structural formula up on the board," Brodie commanded.

Axelrod did so.

"You know," said Brodie as he scrutinized the structure on the blackboard, "if one takes any chemical into the body it's transformed."

Axelrod *hadn't* known that, or at least hadn't known it as acutely as he did now that Brodie had said it. It wasn't just that a drug did something to the body, Brodie was saying, *the body did something to the drug*. It was a revelation.

"What kind of a compound could be transformed to a methemoglobin-forming compound?" Brodie wondered out loud. Methemoglobinemia, a failure of the blood's hemoglobin to bind oxygen, is the name of the condition the headache powders caused; methemoglobin is the deficient hemoglobin responsible.

"One possibility," Brodie went on, "is aniline."

It was just that: one possibility. On the other hand, once you set the structural formulas for acetanilide and aniline side by side it's not hard to see them as related. Both are built up from a single benzene ring, a hexagonal configuration of atoms that is about as familiar a shape in an organic chemistry textbook as a sphere or cube is in a child's set of blocks.

Benzene, a colorless liquid first isolated from coal tar in 1845, had been recognized from the first as made up of six hydrogen atoms and six carbon atoms. But how they were arranged had puzzled everyone for some time. Then along came the German chemist Friedrich August Kekulé who, in an insight inspired by a dream of a snake consuming its own tail, imagined a line of six carbon atoms, each with a hydrogen attached, circling back onto itself and forming a six-sided ring.

Though benzene itself is an organic solvent of no particular note, its hexagonal ring structure serves as a building block for a host of organic compounds. Typically, one or more of its carbon-linked hydrogens is replaced by one or another "functional group," a small cluster of atoms that behaves chemically as a unit. One such cluster is a carbon with three hydrogens; this is a methyl group. An amino group is a nitrogen bound to two hydrogens. An acetyl group is a slightly more intricate arrangement of carbon, hydrogen, and oxygen. Almost invariably, cleaving off such a group from a compound, or adding one, or substituting one for another, confers on the new compound different—often strikingly different—chemical properties.

The headache remedy acetanilide is just a benzene ring with an amino group replacing one of the original hydrogens, plus an acetyl group, in turn, grafted to it. Aniline, the compound Brodie suspected was to blame for methemoglobinemia, is simpler yet: acetanilide with the acetyl group sliced off. The familial relationship between the two compounds is plain to see, and Brodie was taking no unduly fanciful leap of the imagination in guessing that acetanilide was being metabolized to aniline.

But *was* acetanilide being metabolized to aniline? If it were, you ought to be able to give a dose of acetanilide, then find evidence of aniline in the blood or urine. To do that, you needed a way to measure aniline—just as Brodie had needed, and found, ways to measure totaquine, and Atabrine, and procaine. Within two weeks Axelrod had worked one out;

it was based, once again, on methods Brodie and Udenfriend had developed during the war. Acetanilide itself, as well as potential metabolites other than aniline, also proved amenable to measurement. The tools were in place.

•

Julius Axelrod takes good care of his old notebooks. The lined looseleaf pages in which he recorded his first real experimental results as a scientist are, thirty-eight years later, still in pristine condition, bound in a heavy manila file marked simply "Acetanilide." They, and the final paper Axelrod wrote jointly with Brodie, show how the two of them closed in on an understanding of acetanilide's fate in the human body.

One obvious experiment was to administer acetanilide and see how much of it showed up in the stool seventy-two hours later. Answer: essentially none. And how much in the urine? Again, almost none. Conclusion, as Brodie and Axelrod put it in their paper: "Almost all the drug underwent metabolic alteration in the body." It was all being changed into something else.

And quickly. They gave acetanilide and then, hour by hour, tracked its concentration in blood plasma. It rose to a maximum in an hour or so, then fell rapidly as it was metabolized; within seven hours it was gone. Into what had it been changed? Into aniline?

They checked for the presence of aniline in the blood and, sure enough, found it. When they gave aniline directly to dogs, the dogs got methemoglobinemia—and the more aniline, the more methemoglobin found in their blood. On November 25, 1946, Axelrod gave himself fifty milligrams of aniline and analysed his blood for methemoglobin; it rose from almost zero to eight percent within an hour and a half. "I turned blue," he remembers. And woozy. Woozy enough and blue enough and bad enough that, as he says, "I dissuaded some of my colleagues from trying it."

The evidence was mounting in irrefutable support of

Brodie's original conjecture: Aniline was responsible for acetanilide's toxic side effects. But was it responsible for its beneficial analgesic effects as well? It didn't have to be. Frequently a drug is metabolized partly into one compound, partly into another.

Indeed, the prevailing suspicion at the time was that p-aminophenol, not aniline, was the metabolite responsible for acetanilide's analgesic effects, as well as for its toxic ones. Another possibility was N-acetyl p-aminophenol. Both are variants of acetanilide in which one or another functional group differs. Once methods for measuring them were in hand, evidence for their presence was sought in urine and in blood plasma.

The p-aminophenol, as it happens, just wasn't there. But the other candidate, N-acetyl p-aminophenol, did show up in the blood, its rise and fall following that of acetanilide itself. And twenty-four hours later, it was showing up, "conjugated" to certain other compounds, in the urine. Was this the pharmacologically active metabolite? It seemed so. When administered directly, by mouth, it possessed as potent an analgesic effect as acetanilide itself. And it displayed none of its toxic side effects.

On the next-to-last page of their joint paper, Brodie and Axelrod outlined, as they had come to see it, acetanilide's metabolism: A tiny, but important, fraction of it was getting its acetyl group lopped off to form aniline; this was the source of the drug's toxic side effects. By far the largest part of it was following a completely separate metabolic path: A hydrogen atom at the opposite end of the benzene ring from acetanilide's acetyl group was getting an oxygen atom attached to it—was getting "hydroxylated"—transforming it into N-acetyl p-aminophenol. It was this new compound, not acetanilide itself and not aniline, that was exerting the analgesic effect.

The implications of their discovery did not escape them: "The latter compound," Brodie and Axelrod wrote, referring to N-acetyl p-aminophenol, "was not attended by the formation of methemoglobin. . . . It is possible, therefore, that it may have distinct advantages over acetanilide as an analgesic, and it may well serve as a starting point for the synthesis of more effective agents."

N-acetyl p-aminophenol, also known as acetaminophen, is today a widely used analgesic: Brodie and Axelrod had discovered Tylenol.

•

That was the beginning. Now it was *real* science for Axelrod, no more mindless measurement. This was research—probing the unknown, making discoveries. Here, hunches counted. There was no book in which to look up the answers because you were writing the book, you were finding the answers. A lot of the time you groped around in a fog of ignorance, feeling for The Path. So you had to be at home with ambiguity and uncertainty—and Axelrod was. "I took to it like a fish to water," he says. "I enjoyed it, and I was good at it."

For years he and Brodie worked together, the seasoned scientist who'd later be dubbed "the father of drug metabolism," and the technician at his side. Over the next few years, they traced the metabolic fate of numerous other drugs: of the analgesics acetophentidin, antipyrine, and aminopyrine; of the anticoagulant dicumurol; of caffeine, theophylline, dibenamine, and methadone.

It's those early papers Axelrod recalls most fondly. "They still get quoted. They're classic," he'll tell you. Even today he keeps them right in his old gray, government-issue desk, and, as he eagerly digs for one of them to show you, you're left feeling that he's not so much boasting about his scientific achievements as recalling, with longing and pride, the magic of those early days, when he was new to research and he worked at the elbow of his mentor, Brodie. Says Axelrod of their early discoveries: "I couldn't have done it without him. And he couldn't have done it without me."

From the moment he met Brodie, he knew he could never

return to his old job; he never did. On the books, he was employed by the Laboratory of Industrial Hygiene. Physically, he worked at Goldwater, under Brodie. Shirley Udenfriend remembers the day he first appeared at Goldwater: "Suddenly this guy arrives, somebody who didn't come through the army [as had many in the malaria project] and wasn't on the wards. We didn't know *who* he was. Nobody told us. There was just this new guy, lab coat flying, a test tube in his hand. I remember him snapping it to see what color would come up," probably as part of the assay for aniline, which depended on formation of a dye.

Axelrod fell right in. People liked him. He was modest, soft-spoken, and, to use the adjective almost universally applied to him, "sweet." In the lab, he was thoroughly competent. He was a new father around this time, the baby was keeping him up at night, and he'd sometimes come to work groggy with fatigue; even so, he was soon established in the eyes of the others as, in the words of Goldwater technician Betty Berger, "a super-technician. He was the one we'd go to with a problem. 'Let's see what Julie says,' we'd say."

Eugene Berger remembers sensing in Axelrod insight that bored beneath the everyday surface of their work. "I'd be so precise and careful. I'd balance and weigh and get it all down right, and Julie, well," he laughs, "he'd just slop it in. But he'd really *see* what was going on."

Plainly, even allowing for the glow that in retrospect surrounds any Nobel laureate, Axelrod was gifted. And Brodie recognized it. The bond developing between the two men became apparent to all. "Obviously, Brodie had singled him out," says Betty Berger. He'd introduce Axelrod around, take him to meet drug company officials. "It was obvious he was pushing him. And Julie liked that."

For Brodie, too, that early time together glows across the decades. "The first three or four years were wonderful," he says. "Of course, for a year," he adds, "Axelrod didn't really understand what he was doing. But after a few weeks or months, I could see he would be very good for science."

With his confidence in his powers growing under the encouragement of his mentor, Axelrod bloomed. "It was an exciting time," he says. "I'll never forget it." He was happy, ebullient, couldn't wait to get in to work.

He, his wife, and their new baby boy lived in Brooklyn, in a brick apartment building on Caton Avenue, set in a neighborhood of modest row houses not far from Prospect Park. Each day at seven-thirty he'd leave the house, walk to the subway station, and descend to the great sweeping curve of track to await the GG train. It was always crowded, and Axelrod would read his *New York Times* standing up, with the paper folded in fourths, the long way, in classic New York straphanger fashion. At the Fourth Avenue station, the train would break briefly into the daylight and there, as it pulled out of the station, Axelrod could look up from his paper to glimpse the Manhattan skyline rising in the distance over Brooklyn.

Many stops later, he would get off at Queens Plaza, near the foot of the Fifty-ninth Street Bridge, and from there take a trolley halfway across, to Welfare Island. Then it was an elevator down to the ground level, a short walk to Goldwater, and on to the day's work.

One day he might be trying to make an assay more sensitive. On another, he might be measuring drug levels in a patient. "7/19/46," he'd record, in ink, in his notebook:

> Subject: Rizzutto (Male)
> Dose: 1.0 gm (orally) acetanilide, 11 AM 7/18
> Blood: Drawn 1, 2, 3, 6, 11 hrs after taking acetanilide
> Urine collected: 7/18–19/46, 11 AM–8 AM, 1300 ml

Then he'd take a ruler and, this time with a pencil, draw a little grid in which to record the data.

He did everything with his own hands. He ordered chemicals, prepared solutions, fine-tuned methods already shown to work passably, redid earlier experiments, drew graphs and tables. And he *thought*. At the time he made these notes, for example—probably late in the summer of 1946—he was thinking about phenylhydroxylamine (which, in his notebook, he'd often abbreviate "ph a"), the metabolite of aniline he and Brodie thought to blame for methemoglobinemia:

Ideas
1. Effect of aniline on white cells in dog
   " of phenylhydroxylamine "
2. To detm ph OH in presence of aniline
3. Treat aniline with acetic anhy
   1 cc acet anhydride + 5 ml sol.
   acidify & add nitrous to reduce to aniline
   & continue [illegible] perhaps this can be
   done [illegible]
4. Try ascorbic acid
5. Try incub of blood after ph a & acetanilide

Sometimes, for several days running, he'd scarcely see Brodie. "I was just a minor player," he recalls of that early period, just a small part of Brodie's growing empire, which by then numbered about half a dozen people. But then, finally, latest data in hand, he'd go in to see him. "This looks promising," he might say. "I'll go on to the next step, OK?"

A curt "Sure, go ahead" was rare. More likely, when the data was good—that is, when it lent unambiguous support to their current line of thinking—Brodie would become animated and soon the two of them would be launched on a discussion of the problem, its possibilities, its wider ramifications.

Brodie "could really fire you up," Axelrod remembers. "He made every experiment seem earth-shattering." He was always cutting through the experimental minutiae and going for the jugular, for the significance, the meaning that lay behind the results, mapping out the next experiment as he went. He could take a fragment of an idea, a stray scrap of data, and see in it something larger and grander. "Just talking to him made you feel you were having great thoughts, creating great science," says Axelrod.

With Brodie you were often stepping out beyond the limits of the known and driving ahead, on the longest of long shots, to the next experiment. True, you were apt to charge up more blind alleys that way. But it sure beat the tedious routine, the plodding lockstep, to which other researchers often resigned themselves. Instead of thinking up reasons to hold off from trying something, Brodie's dictum was, "Oh, let's take a flier on it."

Let's take a flier on it. There was magic in the style of science embodied in that expression, a breathtaking freedom to be wrong. And Julius Axelrod was not alone in being captivated by it. "Everybody [who worked with him] was changed by that man," says Erminio Costa, who first met Brodie later, at the heart institute. "Whether they admit it or not, they all were changed."

"It's not as if all Brodie's students were his clones," says Johns Burns, who was one of them. "But there was a sort of passing down of genes." The "gene" that was passed down? "To really enjoy science. To make it exciting. 'Why waste time on uninteresting problems,' Brodie would say, 'when there are so many interesting ones to do?' "

Brodie had a way of seeing in isolated results the larger conceptual framework. He always sought the broad sweep, the overarching and the fundamental. He had no fondness for details. Nor for statistics as a means of revealing some subtle pattern in otherwise unclear results; if the pattern was so hard to find, he reasoned, maybe it wasn't there *to* find. Better, he felt, to find the irreducible heart of the problem and go after that.

The Brodie style flourished in his infamous all-night dialogues—dialogues Mimo Costa pictures as almost an art form all their own, the medium through which Brodie expressed his creativity—in which ideas were advanced, then reduced to a series of small logical steps, each testable by experiment. This Socratic method drove some crazy, says Elliott Vesell. Because Brodie never stopped with the easy, obvious interpretation, but always pushed for more.

Vesell, who keeps a substantial personal collection of paintings from the Hudson River School, likens Brodie's style to a Joseph Turner painting: Onto the seemingly flat surface of a phenomenon his intellect would throw new light, reveal in it new facets. Nothing was as it first seemed. Brodie felt, in Vesell's words, that "reality is recalcitrant and doesn't lightly reveal its secrets."

Brodie had a *sense* of biology, of how it works, of what was important and what was not. Science for him wasn't a dusty, closed book, a tight little body of fact and theory. Rather it was something glorious with possibility. His mind sought connections across the widest possible range of phenomena. He was ever the student, wide-eyed and wondering, forever posing questions. "They floated out of him," says Vesell, "like melody floated out of Mozart or line out of Picasso."

Even in his early papers, a hint of the Brodie style appears: The methods he devised with Udenfriend during the war, for example, were the outgrowth of narrowly focused work on a specific class of alkaloids and could quite as well have appeared singly and unheralded. Instead, they were grouped together and coupled to a kind of prologue that explained their significance and outlined a broad strategy for their use. That his methods were developed "in connection with an antimalarial screening program" was noted only parenthetically, stuck literally in the middle of a sentence. Indeed, totaquine, Atabrine, SN 7618, or any other reminder of the malaria program appeared nowhere in the title. Brodie preferred a title giving greater scope to his work: "The Estimation of Basic Organic Compounds in Biological Materials."

That's vintage Brodie—rising above the grubby particulars, soaring into the realm of the lofty, the conceptual, and the grand. For him, there was no such thing as a simple, straightforward result. Everything potentially had wider ramifications. So that, however stuck in the mire of experimental tedium you might be, with Brodie it was as if you were wrestling with the gods themselves. And you'd need to be singularly cloddish and unimaginative not to be drawn in by it. For Axelrod, who was neither, working with Brodie was a heady experience indeed.

Brodie was willing to consider any hypothesis, so long as you could test it. His whole approach to science, otherwise so panoramic, was securely grounded in the pipettes and separation funnels, the centrifuges and colorimeters, of the laboratory. Says Park Shore, "If you suggested to him that the blood actually leaves by the veins and returns to the heart through the artery"—directly contrary to established fact—"he'd say to try an experiment to find out." Brodie had learned from George Wallace, he once told an interviewer, "to go to work in the lab, not to go to the library and read other people's frozen concepts."

One time a young associate, Wolfgang Vogel, came to him outlining what Vogel saw as a beautiful theory. Brodie listened, then suggested a simple experiment to test it. Vogel thought it unnecessary, but did it anyway. As the data rolled in, his theory crumbled.

Try it: Go into the lab with some simple, "quick and dirty" experiment that might immediately suggest whether an approach is worth pursuing. Then, if it works, do it carefully, with all the proper controls. Not for Brodie the long-range planning and tedious preparation in which other scientists often reveled, proceeding by one small, deliberate step at a time. Do now what seems most important, Sid Udenfriend remembers learning from Brodie. "Don't take on a problem that represents some five-year grand project."

Of course, that readiness to go into the lab at the drop of a hat placed a premium on having ideas to test in the first place. And Brodie's receptivity to ideas was, it seems, absolute. There was no such thing as a ridiculous one; he'd turn it inside out and shake it until some nugget of insight fell out. He had little tolerance for negativity, for colleagues too quick to reason out logically why something wouldn't work. It was always *try it*.

Leo Gaudette, another veteran of Brodie's lab, recalls his lab chief once reprimanding him for being unduly critical of a colleague's work. You could always find something in the literature that "proved" something wouldn't work, Brodie felt. He didn't want to hear about it. Most of what appeared in the scientific literature, he was fond of saying, was either widely misinterpreted, or just plain wrong. Why, then, pay heed to it? Why prematurely block off possibilities by taking deceptive refuge in established "fact" that, often enough, isn't?

In fact, Brodie was on a less intimate basis with the research literature than some other scientists. Too much knowledge merely inhibited new ideas, he felt; accordingly, he'd try to obliterate from his mind all he knew about a particular topic, pretend he knew nothing. Often he *did* know nothing. "Sometimes he'd shock people by asking obvious questions," says Elliot Vesell. Once, the story goes, when investigators in his own lab discovered an important class of enzymes in the part of the liver called the microsomes, Brodie kept calling them "mitrosomes," a source of hilarity around the lab for some time.

But Brodie simply saw no reason to become an expert in an area to launch a study of it. Rather, as Sid Udenfriend says, "he would just wander into a new field and make advances that people fifteen years in the field couldn't." Poring through scientific journals didn't appeal to him; picking the brains of colleagues did. "He'd go up to you," Jack Orloff remembers, "and say, 'Tell me what you know about X and Y.' Sometimes he'd already know a lot, but he could come across as almost stupid." Indeed, he could seem downright ignorant, asking disarmingly simple, even hopelessly naive questions, like a child. But as one admirer notes, "He'd end up asking just the questions you should have asked ten years ago."

One who was a member of Brodie's group later, James Wyngaarden, describes Brodie's mind as "like a bear trap. He remembered everything you told him. He'd tour the lab with a foreign visitor and go from one to the other of us, filling in what we each were doing, down to the last detail. It was mind-boggling."

One time, Brodie and he got to talking about some thyroid-related work Wyngaarden had begun in another lab. Brodie knew nothing about the thyroid, and at the beginning of their conversation asked the most elementary questions. But imperceptibly they deepened, growing in complexity, and by the end of their talk he was asking questions that Wyngaarden found penetrating and provocative. "I was very surprised," he says, "because fifteen minutes before he knew nothing."

In any case, notes Jack Orloff with a shrug, Brodie's approach served him well. "For years everything he touched [scientifically] was gold." A man trained in organic chemistry and almost wholly ignorant of biology, he ultimately made important contributions to drug metabolism, physiology, neurochemistry, even genetics and evolution.

"He was an individual, a character. He had genius," says John Burns. When the two first met in 1950, Brodie impressed him greatly with his lively mind. "Everything he said was so exciting." And that, in the end, was what changed the lives of so many of those he touched: Brodie was *exciting*. His ideas were exciting. The way he couched them was exciting. Everything he touched seemed to burst with vitality.

"I owe him a great deal," says Julius Axelrod. "More than any single individual," he says, "he started me off on a research career."

•

But Axelrod could go only so far with Brodie. There was another, darker side to working with the man. Brodie was infinitely receptive to ideas. But as to which of them to follow up, he called the shots. You couldn't just go off and follow your scientific nose, as you could in looser labs. He was brilliant and inspiring, but also dictatorial. "This is what you're going to look for, and this is what you're going to find," is how one colleague remembers him operating. Brodie's lab was not always a happy place; there could always be heard plenty of bitching, with "graduates" often coming to appreciate its impact on them only later. "Brodie picked the problem," remembers Herb Weissbach of the early heart institute period. " 'This is the way the experiment is done,' he'd say. 'You do this, and you do that.' "

At Goldwater, "Brodie laid down the responsibilities on Julie's head, which he was prone to do," recalls Gene Berger, "and Julie accepted them, which *he* was prone to do." But though grateful for how Brodie had opened up new vistas for him, in time Axelrod wanted a chance to explore them on his own. He was still, to Brodie, a super-technician. His master's degree counted for little in a field where the ticket of admission was a Ph.D. He was still under Brodie's thumb.

However closely they worked, their relationship remained that of lab chief and technician, period. Brodie was professionally friendly, even charming. He was a superb raconteur, told great jokes, and regaled his associates with stories of his poker exploits.

But that was as far as it went. Brodie ate in the doctors' dining room upstairs; Axelrod, a mere technician, did not. Brodie demanded respect. He did not encourage familiarity. "Even Julie called him *Dr. Brodie*, never Steve," says Gene Berger. Brodie displayed confidence that verged on imperiousness. "He was smart, smarter than everyone," says Axelrod. "And he showed it." Even to the Goldwater M.D.s, not accustomed to listening to Ph.D.s or anyone else for that matter, Brodie gave orders. The lab was Brodie's show—and Axelrod wanted a show of his own.

Then one day—it was April 7, 1949—Axelrod opened the *New York Times* and saw it: "Appointed Research Head of Heart Institute," the one-column headline read. "Dr. James A. Shannon, director of the Squibb Institute for Medical Research, New Brunswick, N.J., has been appointed associate director of the National Heart Institute in charge of research, the Public Health Service announced today."

This, thought Axelrod, was his chance.

5.
Building 3: "All He Had to Do Was Whistle"

THERE WASN'T MUCH to see at NIH in 1949. Today, the National Institutes of Health has its own underground station on the Washington, D.C., Metro. The largest medical library in the world is here. From along busy Rockville Pike, flanking NIH, you can glimpse the great laboratory complexes, the squat animal quarters, the 1.3-million-square-foot Clinical Center, with its alternating corridors of research labs and medical wards, the office towers and power stations and parking structures that sprawl across this 308-acre campus.

But back in 1949, when Julius Axelrod read in the *New York Times* that James Shannon had been named scientific director of the heart institute (part of NIH), NIH was still mostly countryside—the verdant, gently rolling countryside of Montgomery County, Maryland, twelve miles northwest of downtown Washington, just outside a town that, by the 1940 census, was home to barely two hundred souls. After a lifetime in bustling New York City, this was where Julius Axelrod pinned his hopes and his future.

He'd met Jim Shannon only once, and then but briefly. To him, Shannon was a legend: leader of the Goldwater malaria project, force behind its creative excellence, the man who had discovered Steve Brodie. Propelled by the vision of a scientific life apart from Brodie, he wrote Shannon, set up an interview, went to see him at Squibb, and wound up with the promise of a job. He was bound for Bethesda.

He didn't know that half of Shannon's old group at Goldwater, Brodie among them, would be bound for Bethesda with him.

•

NIH goes back to a one-room bacteriological laboratory established on Staten Island, New York City's least urban borough, in 1887. One of the lab's first major projects was the study of cholera and infectious diseases among the eastern and southern European immigrants then coming to America in record numbers.

In 1891, what had become known as the Hygienic Laboratory moved to Washington, D.C. By 1930, when Herbert Hoover signed legislation transforming it into the National Institute—not yet plural—of Health, the Hygienic Laboratory, located on a five-acre tract at Twenty-fifth and E streets in northwest Washington, just down the street from a brewery, consisted of a pair of modest two-story buildings.

In 1935, part of a ninety-two-acre Montgomery County estate, "Tree Tops," was donated to NIH. Visions of microbial invasions and laboratory animals running wild set off alarms among area civic groups, but Gilbert Grosvenor, editor of *National Geographic*, whose property lay just north of the future NIH site, helped quell the uprising. Construction of Building 1, the NIH administration building, began in January 1938, as did that for two other modestly-scaled lab buildings: Building 2 was dedicated to the study of industrial hygiene, while Building 3 was named the "Public Health Methods and Animal Unit Building."

But bigger things were in store for Building 3. Within a few years of James Shannon's appointment, it had become one of the most fertile research settings in the world. "I've never been in a place where the concentration of talent was so high," says Herbert Weissbach, now director of Roche Institute of Molecular Biology in Nutley, New Jersey. "It was an incredible experience. . . . Building 3 was never to happen again in my career."

During that period, three future NIH directors, two future Nobel laureates, nine future members of the National Academy of Sciences (almost *one per cent* of its current membership) worked in a structure about the size of a big elementary school. What accounted for the Building 3 phenomenon? Julius Axelrod was asked as he sat at lunch in an NIH cafeteria years later. "*Shannon!*" he exclaimed, the soft *sh* sound rolling around in his throat like a Scotsman's brogue. "He's the one who did it."

Shannon had left Goldwater in 1946, to become director of the Squibb Institute for Medical Research. There, he helped recognize the potential of a new antibiotic, streptomycin, and saw to its expanded production. But, says Dewitt Stetten, who first met him at Bellevue Hospital back in the 1930s and served under him at NIH for eight years, Shannon soon grew disenchanted with the pharmaceuticals industry. Stetten pictures Shannon as an austere man of old-fashioned morality, devoted to his family, who never was heard to laugh at a dirty joke, and who adhered firmly to the tenets of his Roman Catholic faith. At Squibb, he says, Shannon felt tainted by "too much of the high life, by too much money." That, and seeing free research unduly distorted by commercial considerations, finally forced him to leave.

By 1949, Shannon was science director of the National Heart Institute and was embarked on a great mission: In the belly of the federal *bureaucracy*, for God's sake, he was going to make a topflight research center. It was Goldwater all over again, only bigger. Deliberately, systematically, he set about building up a scientific team. "He had a lot of contacts in science," says Shannon's old friend Tom Kennedy. "He went to those he respected and got lists of the best people. When the same name got mentioned two or three times, he'd go after him. . . . All he had to do was whistle, and people came running."

Well, not quite. Today, with the early 1950s at NIH blurring together into a single "early period," it can seem that way, but it wasn't so easy. True, in Shannon's favor was the paucity of good positions then available for young scientists. Another factor was the Korean War: A job at the Public Health Service, with duties divided between ward and laboratory, kept a bright young doctor out of the army. One could satisfy one's military obligation, the law held, by membership in any "uniformed service"—of which the Public Health Service was one—not just in the armed services.

James Wyngaarden, the current NIH director, tells the story of how, eligible for the Korean draft despite his service in World War II, he was at one point "invited" to accept a stint in the army as a medical officer, or else be drafted. He had twenty-four hours to make up his mind. Shannon, who'd met Wyngaarden on a recruiting expedition to Harvard, wanted him for the heart institute. "Sit tight," Wyngaarden remembers Shannon telling him. "Don't sign anything." At eleven o'clock that night, Wyngaarden's commission in the Public Health Service came by telegram.

But working against Shannon was widespread skepticism that government research could breed anything but mediocrity. "It's going to be the most gigantic backwater you ever saw," a distinguished Harvard professor warned Donald S. Frederickson, still another future NIH director, as he was about to move to Bethesda in 1952. That sentiment was a prevalent one.

When Sidney Udenfriend heard from Shannon in 1950, he was in Saint Louis, on a postdoctoral fellowship in the Washington University laboratory of Carl Cori, who'd won a Nobel Prize a few years earlier; Earl Sutherland, a future Nobelist, worked down the hall. In his letter, Shannon explained how he wanted to build up the heart institute and asked Udenfriend, thirty-two at the time, to come help him do it. Udenfriend asked Cori, his lab chief, what he thought.

" 'Who wants to work for the government?' " Udenfriend remembers Cori replying. "His idea of me going to NIH was that I was ending my career," says the founding director of the Roche Institute of Molecular Biology. "To him, government science was like the National Bureau of Standards, or the Department of Agriculture." Udenfriend had applied for an assistant professor's position at Columbia University, but the school was taking its time about responding. Shannon pressed him for a decision, and finally Udenfriend opted to join him in Bethesda. "It was definitely a second best," he says. "I took it by default."

When Bob Berliner heard from Shannon, he, too, at first declined his old boss's invitation; his mentor at Columbia was suspicious, in principle, of any government role in research. "Why don't you at least come down to talk?" Shannon importuned him. Berliner remembers the scorching midsummer day on which he did. In the end, he told Shannon he wasn't coming. Except it *wasn't* the end. Shannon persevered. No, said Berliner. Yes, said Shannon. Yes, said Berliner.

A little later, Gordon Zubrod, another ex-Goldwater man, came from Saint Louis University. From Goldwater, to which he'd returned after finishing his medical residency, came Tom Kennedy, who took a position under Berliner in the heart institute's Laboratory of Kidney and Electrolyte Metabolism. Robert Bowman, an M.D. who'd made a name for himself at Goldwater with his knack for coming up with ingenious lab instrumentation, also came down from New York. And so, in the end, did Steve Brodie.

Axelrod remembers Shannon showing up at Goldwater and spending two days in Brodie's office, the door shut, presumably trying to prevail upon him to make the move. Brodie's friend Joseph Post remembers once talking with him about it all afternoon. "It was a chancy thing," says Post. "It was the government. No one knew how long it [the heart institute experiment] would go on. No one knew whether Shannon could do it or not."

Ultimately, of course, Brodie figured Shannon *could* do

it, and joined the others in Bethesda. Explains Tom Kennedy, "At Goldwater, Brodie was master of three hundred net square feet of lab space. Here comes an offer that gives him fifteen times as much and the opportunity to have three or four section chiefs under him." How could he *not* take it?

And Axelrod? In approaching Shannon for a job, the idea had been to get out from under Brodie and go his own scientific way. He got the job, all right—but as a technician for Brodie! He had a new position, in a new town, on the ground floor of an exciting new scientific enterprise. But he was still deep in Brodie's shadow.

•

Among Shannon's recruits were Christian Anfinsen, who'd worked on the Harvard end of the malaria program during the war and now found his assistant professor's salary doubled to today's equivalent of about forty thousand dollars. Shannon, says Anfinsen, was "a real high-powered guy," who picked people based not on their formal credentials alone but on their capacity to spark excitement among their colleagues, and so enhance the research ambience as a whole.

And that's what happened in Building 3, where Shannon's new crop of heart institute researchers were concentrated. Back at Harvard, there'd been "too many suits and jackets" for Anfinsen's taste. But Building 3, a three-story brick structure whose slate roof, dormers, arched doorways, and elegant Georgian detailing made it look like a college dorm, was a real workplace. "We were crammed in there, ten of us in one lab. But we were so obsessed with the work itself, we didn't realize it was so crowded. . . . There was a sense of familyhood." They were young, eager, bursting with ideas.

Sometimes, on a pretty day, a few of them would sit out on the lawn in front of Building 3 and talk science for hours at a time. Then each week there were lunch seminars, where one or another of them would present a recent paper of interest; a list was posted and you'd just run your finger down it to find when it was your turn. There were also larger biweekly or monthly research talks, drawing forty or fifty at a time, held in a temporary building nearby.

Donald S. Frederickson would later reminisce how back at Harvard, he'd rubbed shoulders with many top researchers, some of them true giants. "But there wasn't anything like the number of giants we encountered easily and frequently in Bethesda." You could go into the cold room on the first floor of Building 3, where chemicals and biological materials were stored, and there encounter a whole succession of bright young people, most of them destined for distinguished careers, eager to offer advice and criticism. "The critical mass was there," says Frederickson. "It was said that after encountering some strange disease on morning rounds, [an M.D.] should have thought of the affected enzyme by noon, be in the laboratory of an expert on that enzyme by three, and be ready to discuss one's protocol to test for the deficiency at next morning's rounds."

Frederickson, Anfinsen, and the others from Harvard were clustered on the first floor and down in the basement. "We felt rather like strangers in this crowd from Goldwater Memorial Hospital," remembers Frederickson. The Goldwater contingent was spread over the second floor and part of the third. It was a stimulating group, recalls Bob Berliner. "We had a wonderful time."

Tom Kennedy and Julius Axelrod first reported to work at NIH in December 1949. The two of them had seen each other around Goldwater, and arranged to come down together. Kennedy drove. Naturally, being New York natives marooned in the wilds of Maryland, on that first day they got lost.

The two of them found apartments in the same complex in Silver Spring. Later, Sid Udenfriend and his wife joined them there. The Axelrods had had a flood in their basement apartment, Udenfriend recalls, yet helped them get settled in theirs. "For a year or two, we were very close, even family-wise." But then, as one or another of the Goldwater crew settled into the area and bought houses, they drifted apart, though they saw each other daily at the lab.

During this period there was a regular poker game among the Building 3 group comprising, among others, Anfinsen, Udenfriend, Berliner, and Brodie. Each week the locale shifted. "There will be a meeting of the Association of Applied Statistics at Sid Udenfriend's house on Friday night," a notice posted around the lab might read. Everybody knew what it meant. The game was a solid enough fixture of their Friday evenings between about 1952 and 1956 that, says Udenfriend, his wife bought a special card table, the kind with recesses to hold drinks, that was hauled out just for their games. Udenfriend remembers Bob Berliner as the statistician of the group, whereas "Brodie relied on intuition and guts." Most always, Brodie won.

For Brodie, who'd recently married, uprooting himself from New York City didn't come as easily as it had for some of the others. His first apartment was in Pook's Hill, a suburb even further outside downtown Washington than NIH itself. It didn't take. "If I have to wake up one more morning to singing birds, I'll blow my brains out," Brodie could be heard to grumble. Soon, he and his new wife moved to the Statehouse Apartments, near DuPont Circle, a new, nine-story brick edifice across Massachusetts Avenue from an exclusive private club.

Brodie's résumé paints him as severing his Goldwater ties in 1950 to become head of the Laboratory of Chemical Pharmacology at the National Heart Institute. In fact, for several years he continued to run Goldwater as a kind of sideline, periodically commuting between Bethesda and New York. His right-hand man at Goldwater, overseeing it while he wasn't there himself, was John Burns, to whom he was introduced in 1950 while at a scientific meeting in Atlantic City.

Burns had first heard of Brodie during the war, when he'd been assigned to work in the malaria program at Atlanta Penitentiary. In New York, Brodie took the twenty-nine-year-old Columbia University doctoral student under his wing. Soon after they met, Burns remembers, Brodie brought him to a pharmacology conference in Boston and, on the New Haven Railroad train bound there, broke out a copy of Goodman and

Gilman, a standard text, and began tutoring him in pharmacology. By the time they reached Boston, Burns knew enough to follow what was going on at the abstract sessions (where brief summaries of upcoming papers are presented). "This," he smiles today, "was my first experience in pharmacology."

Burns had planned to do a postdoc in biochemistry at Columbia. But talking with Brodie changed all that and soon he was working in Brodie's lab on drug research. For ten years he hopped between New York and Bethesda. And until about 1954, Brodie, in turn, often came up to New York to confer with Burns, review current research, and generally survey his domain.

•

Lewis Aronow remembers wending through corridors to the basement and there, in a small, crowded, equipment-packed office, meeting Steve Brodie for the first time.

It was the spring of 1950 and Aronow had just gotten out of CCNY with a bachelor's degree in chemistry. He didn't know much about medical research, he says, but *did* know he didn't want to go to medical school; he felt uncomfortable around sick people. So when a friend advised him that Brodie was looking for lab technicians, he arranged for an interview.

Brodie impressed him as charming, knowledgeable, and enthusiastic. The lab's approach to pharmacology, Brodie told him, depended on gaining knowledge of blood levels, whether in animals or man; from that, all else followed. As for Aronow, he almost literally didn't know how to *spell* pharmacology, much less what was so special about blood levels. He'd done well enough at City College, though, had good recommendations, and got the job.

But the job was not at Goldwater, where Brodie had interviewed him. Rather, he was hired at the federal civil service grade GS 5, and told to report to Building 3 of the National Heart Institute in Bethesda, Maryland. He did. "I walked into a lab and there was a man unpacking a box of pipettes and loading them into a drawer. It was Julie."

The lab was still being set up. Everything was brand new. Axelrod himself hadn't been there long. As for Brodie, he remained mostly up at Goldwater, coming down to Bethesda only once every week or so. So it was Axelrod who set up Aronow with a project. Aronow found him unassuming—"about the least intimidating man I've ever known"—and a pleasure to work with. "He'd never ask you to go beyond what was normal, and yet you felt you had to keep up with him."

At the lab bench, Axelrod was a whirlwind. "He just *poured* the data out," says Herb Weissbach, who joined the lab a little later. "Problems did not beat Julie. When an experiment didn't work out, Julie figured out a way in which it *would* work out." Weissbach's most vivid image of him? A man with a patch over one eye and a cigarette dangling from his lips, wildly running around, mixing, weighing, washing, setting up equipment. One felt loath to disturb him; he seemed driven by some relentless internal gyroscope.

One time, an experiment had Axelrod running from one end of the small lab to the other, doggedly oblivious to his surroundings. A centrifuge stuck out into the aisle along which Axelrod repeatedly passed and, on a lark, someone pushed it an inch or two further into his path. Next time through, Axelrod didn't notice. Over the course of the day, the centrifuge was edged further into the aisle, the gap between it and the opposite lab bench gradually narrowing. And still Axelrod never noticed, though by the end of the day he was reduced to twisting and wriggling his way through.

Axelrod kept scrupulously regular hours. But while in the lab, he never stopped. "I've never seen anyone who wasted less time than Julie," says James Wyngaarden, who for two years during the early fifties car pooled with him. That car pool, he recalls fondly, was "like a moving seminar, dominated by Julie and Gordon." Gordon Tomkins, now dead, was a young biochemist blessed, Wyngaarden says, with insatiable curiosity and an ability to recall obscure biochemical details. Axelrod was forever peppering him with questions: "Did you read the paper on such and such?" Tomkins always had.

"Well, doesn't that disagree with X?" "Yes," Tomkins would reply, "but I think Aay did this, while Bee used a different buffer."

"Most of us," says Wyngaarden, "skim lightly over details like that. But these guys picked up every nuance." The talk might touch on any topic, from the gritty details of a particular experiment to Linus Pauling's triple-helical model of DNA, at the time not yet displaced by the Watson and Crick double helix. There was little talk of people or personalities. It was all science, with Axelrod asking questions from the moment he got into the car.

Joining Brodie's lab during this period was Park Shore. Shore had gotten his bachelor's degree in chemistry from George Washington University, flirted with radiation chemistry as a career, decided it was a dead end, and, casting about for something else, heard about a new group at the heart institute that emphasized a chemical approach to pharmacology. Sounds interesting, he thought, and arranged the interview with Brodie that, as he says, "completely changed my career."

Like Aronow, Shore found himself working not with Brodie, who was still mostly up in New York, but Axelrod. "He was Steve's deputy, his *el segundo*." It was Axelrod who showed him around the lab, gave him his first project, offered advice. Shore found him animated, easy to talk to, and thoroughly likable.

At first, Axelrod seemed simply bound up in his work. "But it wasn't long before you could tell he felt things were a little one-sided" in his relationship with Brodie, says Shore. "To me and others in the lab, it was amazing he put up with it." Axelrod was as talented as anyone there. Yet while other technicians were breaking free from Brodie and becoming independent scientists, he seemed immobilized, remaining merely "Brodie's technician."

The way Shore tells it, Brodie viewed anyone in the lab as a sort of postdoctoral fellow, still in training, there to do the lab chief's bidding, not run off on projects of his own.

When Udenfriend rejoined Brodie at the heart institute, this time with a Ph.D. in hand, he took care to reach an understanding with him from the start. "I put it straight to him," he says. " 'You carry out your work, I'll carry out mine.' " But gaining his professional freedom took a forceful and unambiguous statement of will.

"Many of us told Julie he was crazy to stay there," says Lewis Aronow, who after two years with Brodie went off to Harvard for his Ph.D. "But it was hard for Julie to be assertive. He had considerable self-doubt." Not that he was scientifically diffident. Even back at Goldwater, remembers one who worked with him there, he had plenty of confidence in his abilities. But now family and personal responsibilities had him hamstrung. And besides, he knew independence would mean a final break from Brodie.

Axelrod was not alone in finding it hard to break free from Brodie. "I had that problem early on," says Gene Berger, Brodie's technician at Goldwater. "It's like a child leaving a parent." When Brodie left for NIH, Berger reports that in some ways he felt relieved at not being asked to join him. "I wanted to be independent." But if Brodie *had* asked him? "A difficult question," he replies. "I can't answer that even today."

Brodie was apt to take it as a personal affront when a trusted subject left the realm. Mimo Costa tells of driving Brodie home from work one evening in the mid-1960s and abruptly announcing that, after six years, he was taking a job at Columbia the following month. Brodie was hurt. "You should not have done that," Brodie said, and got out of the car. Why didn't Costa talk to him about it beforehand? "Because," he says, "I knew that if I had he'd talk me out of it."

Getting out from Brodie and going out on his own, Axelrod knew, meant going for his Ph.D. As far back as 1951, Sid Udenfriend remembers, he and Axelrod had talked about graduate school. The topic came up several times at lunch, once or twice when their families were together. But, he says, Axelrod threw up all sorts of practical reasons for why he felt he couldn't do it. He was already making a good salary. He loved the work. ("He was dissatisfied with The System, but always happy around the lab.") And how could he take time out for grad school with a wife and two kids to support? Besides, the last time he'd been in school was 1941. He'd be going back, at age forty or so, to exams and all the other adolescent trappings of school.

In time, though, as Herb Weissbach says, "Julie saw that being Brodie's technician forever, until he retired or until he died, was not something he was comfortable with." There was already a vehicle, of sorts, for him to return to school: Brodie had started an "underground graduate school," to use Udenfriend's expression, where many of his technicians went for doctorates at George Washington or Georgetown universities, all the while holding down jobs at NIH. It was not, apparently, strictly legal, but many Brodie hands were doing it. At least once or twice, Axelrod brought up the possibility to Brodie. "He never told me just what Brodie said," says Udenfriend. "All I knew was, he wasn't doing it. Brodie was not encouraging him."

He didn't encourage me. To this day, Axelrod gives this as the main reason why for so long he didn't go after a doctorate. Oh, it wasn't all Brodie's doing, he concedes. Some part of him just didn't want it enough. And, truth to tell, he had a nagging fear of the foreign language requirements. Still, thirty years later, what he views as Brodie's failure to push him still rankles. "If he'd encouraged me," he says, "I would have gone after it. He never said I was good, that I wouldn't have any problems with it."

Brodie denies blocking Axelrod's way. Even back at Goldwater, "after a year's working with him I would have said he would make good doctoral material. At NIH, my technicians were all going to school. He was the only one who wasn't. I didn't hold him up."

But some who saw the relationship between the two unfold believe Brodie simply didn't want to let Axelrod go. "Brodie was very dependent on Julie," says Lewis Aronow. "He honestly thought it was an ideal relationship, with him the idea man, and Julie the absolutely perfect man for coming up with the data." Plainly, Axelrod was a rare and supremely valuable resource. And just as some feel Brodie made Axelrod as a scientist, there are those who believe that, without Axelrod, Brodie's discoveries would have been fewer, his stamp on science less pronounced.

Axelrod had tried to break from Brodie in coming to NIH, and now, well into the 1950s, was with him still. Others, like Shore and Aronow and Jack Cooper, were coming into the lab with bachelor's degrees, getting encouragement from Brodie to go for their doctorates, doing so, then going off on their own scientifically. But not him. Why didn't Brodie encourage *him*?

Says one of Axelrod's old colleagues, "Julie was too clever a guy to sit on the sidelines forever." But it took the microsomal enzymes discovery to propel him into The Game at last.

•

Where you begin the microsomal enzymes story inevitably colors where it ends—that is, with whom credit for their discovery properly lies. Do you properly begin it with the day Steve Brodie got a phone call from a pharmaceutical company about a strange new compound that seemed to enhance the effects of other drugs, yet by itself did nothing? Or does the story more appropriately begin when Julius Axelrod first started reading about a group of compounds called the sympathomimetic amines and decided to trace their metabolism?

More than two thousand years ago, Chinese physicians began grinding up the herb they called *ma huang* and using it as a cough remedy, fever reducer, and circulatory stimulant. In 1887, *ma huang*'s active principle was isolated and named ephedrine, after *Ephedra*, the genus to which *ma huang* belongs. Ephedrine was later found to be one of a class of compounds, similar in structure and behavioral effects, each able to mimic, to a greater or lesser degree, the body's sympathetic nervous system. Because chemically they all shared an amine group, they were dubbed the sympathomimetic amines. Amphetamine, known on the street as speed, and mescaline, the hallucinogenic agent in peyote, are also members of this family.

In 1952, Axelrod began studying ephedrine and amphetamine, about which virtually nothing then was known. Getting a green light from Brodie to pursue the problem largely on his own, within a year he had traced the general features of their metabolism. That is, he had devised means of measuring them and their metabolites and had outlined their metabolic pathways. He found, for example, that ephedrine was in some animal species being demethylated, and in others, hydroxylated. (To be demethylated means simply that a compound is transformed into another by the loss of a methyl group. To be hydroxylated means the transformation occurs through the gain of a hydroxyl group.) The fifteenth paper of Axelrod's scientific career was called "The Biotransformation and Physiological Disposition of L-Ephedrine and L-Norephedrine," and for the first time, Brodie's name did not appear as co-author.

At this point, Axelrod's work was still a species of Brodie's New Pharmacology. But now he was after more: How did these metabolic changes come about? What enzymes were responsible and where could they be found?

An enzyme is a biological catalyst, urging a reaction forward without itself taking part. That, at least, is the traditional definition; in fact, it hides the truth. Because enzymes typically speed things up by a factor of ten million or more, it is fair to say that, without them, biochemical processes that can occur, don't. In short, enzymes are essential to life.

Which enzymes, Axelrod wondered, are responsible for metabolizing ephedrine and amphetamine?

•

One day in the early 1950s, Steve Brodie got a phone call from Glenn Ullyot at Smith, Kline and French Laboratories in Philadelphia. A compound the company called SKF 525-A had been found to exert a peculiar effect when given along with other drugs. Administer SKF 525-A by itself and, at all but very high dosages, it did nothing. But give it along with, say, a barbiturate and the barbiturate became more of a barbiturate; its action was prolonged. Give it with narcotic analgesics like morphine or codeine, and those drugs worked longer. Give it with amphetamine and its effect was prolonged as well.

If SKF 525-A was enhancing the potency of drugs each so different from one another, perhaps it was retarding their metabolic destruction through some common mechanism. Or, as Brodie put it in a review of the subject a few years later, "It seemed to us that if a number of drug metabolic pathways were susceptible to the same inhibitor they should have certain factors in common."

What were these factors? Soon, several members of Brodie's lab were at work on various aspects of the problem.

•

Axelrod was doing something he had never done before. He wanted to know where in the body amphetamine was being metabolized. More specifically, where were the enzymes that did it? He was tracking an enzyme, yet he was no enzymologist. But in Building 3 was his friend Gordon Tomkins, who was.

"You know, there's no big mystery to being an enzymologist," Tomkins advised him. "All you need is a razor blade and a liver." The liver, because it was the site of so many enzymatic processes in the body. A razor blade to dice it up.

"Have you got a method for amphetamine?" Tomkins went on. He meant, did Axelrod have a ready means of measuring it? He had. It was based on the methyl orange reaction Brodie and Udenfriend had cooked up a decade earlier. "Well, why don't you throw amphetamine in a liver slice?"

If amphetamine simply disappeared when you did what

Tomkins suggested, it meant that it was being metabolized into something else. Sure enough, it did disappear; it was, Axelrod found, being deaminated, or losing an amino group, and so becoming phenylacetone. But where in the liver were the enzymes responsible for this transformation?

Enter the ultracentrifuge.

A centrifuge is a standard piece of laboratory hardware used for separating an outwardly homogeneous sample into its components. By rapidly spinning a test tube, not around its own axis but as if it were the spoke of a wheel, denser parts gravitate toward the bottom of the tube, forming a pellet. Lighter ones remain in a supernatant (meaning "floating above") layer at the top. The new ultracentrifuges spun faster—up to one hundred thousand revolutions per minute, compared to the fifteen thousand r.p.m. or so then common—thus permitting finer separation.

This technological advance invited a whole new experimental strategy, differential ultracentrifugation: After finely grinding up the tissue in question, an initial low-speed spin separates out the heavier cell components. In subsequent spins, these heavier parts can themselves be separated one from the other. The original supernatant, meanwhile, is further spun down, this time separating out from it the lighter cell components. And so on. "It was the thing, then, to see what part of the cell" an enzyme came from, Park Shore remembers of the period. Is it in the nucleus of the cell? The cell membrane? You'd spin down each, then see if that "fraction" retained enzymatic activity.

This was the strategy upon which Axelrod embarked. He had already found that if amphetamine were added to ground up rabbit liver, the amphetamine disappeared; it was metabolized. Now he simply repeated the experiment using liver cell fractions he had isolated by differential ultracentrifugation. Did the cell nuclei, under appropriate biochemical conditions, metabolize the drug? No, they did not. What about the mitochondria? No. The microsomes? No. The soluble supernatant fraction, which is what's left after all the centrifuging? No. What about combining the nuclei and the supernatant? Still no. Or the microsomes combined with the supernatant? *Yes.*

The microsomal fraction and something in the supernatant were together able to metabolize amphetamine. But which of the two contained the enzymes? And what did the one without the enzymes contribute?

To answer the first question, Axelrod hit upon the kind of elegantly simple experiment for which he would become famous: Certain enzymes work only at body temperature. Raise the temperature much above thirty-seven degrees Celsius (ninety-eight degrees Fahrenheit) and they no longer function. What if, Axelrod wondered, you heat the supernatant and keep the microsomes at body temperature, and see whether the two together continue to work? Then reverse the procedure: Heat the microsomal fraction and keep the supernatant at body temperature, and see what happens then.

Sure enough, when the microsomes were heated to fifty-five degrees Celsius for ten minutes, the amphetamine was no longer broken down: The enzymes *had* to be in the microsomes!

And the "something else" in the supernatant that made the whole process work? Axelrod knew that a kind of enzyme helper known as TPN was crucial to certain enzymatic processes. Yet even with TPN present, as he'd found earlier, microsomes alone did not metabolize amphetamine. The recipe needed the supernatant. Working hypothesis? Something in that supernatant transformed TPN into another compound—a compound the microsomal enzymes needed in order to work.

The final piece of the puzzle slipped into place through the help of a colleague, Bernard Horecker, who'd been working with enzymes whose functioning required TPN. Horecker supplied Axelrod with certain key substrates (the material on which an enzyme acts). And any one of them, Axelrod found, added to TPN and microsomes, metabolized amphetamine.

"These substrates had one thing in common," he wrote later. "They generated TPNH," what chemists term a "reduced" form of TPN.

Apparently, it was this latter compound that the microsomal enzymes really needed: the supernatant furnished enzymes—ones not so sensitive to heat as the microsomal enzymes were—which turned TPN into TPNH. As Axelrod by now expected, when he synthesized TPNH and "fed" it to microsomes, the amphetamine was duly metabolized.

By the end of June 1953, Axelrod had the details worked out. Later, he subjected another drug, ephedrine, to a similar round of experiments and got identical results. Though ephedrine was metabolized by a completely different biochemical route—by demethylation, rather than deamination—it seemed that the same enzymes, in just the same location, needing the same cofactors, were responsible for bringing it about.

Axelrod presented the amphetamine results before a small audience at the fall 1953 meeting of the American Society of Pharmacology and Experimental Therapeutics. A one-paragraph abstract, or preview of a coming paper, appeared in the society's journal the following year. It contained no hint of any larger significance. But the paper that followed did: "It is becoming increasingly evident that enzymes in liver microsomes which have a specific requirement for reduced TPN and oxygen are of major importance in the detoxification of many drugs and foreign organic compounds."

•

Spurred by SKF 525-A's promise of a common factor in metabolizing drugs, Brodie had put his whole lab to work on the problem. "At first it was believed that the various reactions would be catalyzed by completely different enzymes," James R. Gillette, who joined Brodie's lab during this period, later recalled. And so, various drugs had been assigned to various investigators.

"They were very systematic," recalls Axelrod, "just plodding right along," when along came his microsomal enzymes abstract. Axelrod, agrees Steve Brodie, "was working on his own and frankly we did not know what he was doing until we opened up the [abstract] paper and saw it." Soon, his results were confirmed by the others and extended to other drugs. "Once I had got it," says Axelrod, "it was easy for them to continue my work."

Plainly, microsomal enzymes were big news, the kind of important phenomenon, soaring across normal conceptual barriers, that was right up Brodie's alley: The whole idea of an enzyme is its riflelike specificity, that it acts on one or a very few specific substrates. Yet here was an enzyme system that was more analagous to a shotgun, that seemed able to handle a whole host of drugs. *This was nothing less than the biochemical system nature supplied for the detoxification of foreign substances!* As they scraped for berries or bark or insects, animals ingested all sorts of foreign compounds. The body needed a way of handling, not just drugs, but all such compounds, many of which it had never "seen" before. This was it.

It was a discovery warranting far more attention than that generated by Axelrod's little abstract, or even by the longer article due to follow it. Then, too, as Brodie reconstructs it, "My lab was up in arms." Axelrod, working on his own, had walked off with a prize they'd all sought. "I got them all together," says Brodie, and suggested a joint paper, in an important journal, with all of them as coauthors. That way it would have far more impact. "It was the only thing I could do," he says.

Axelrod remembers the meeting, too, his mouth tightening as he does. "Let's all publish together," he recalls Brodie saying. "We'll all go in alphabetical order. . . ." And in that pause, Axelrod imputes to Brodie the abrupt realization that alphabetical order meant Axelrod would appear first. "Except," Brodie then added, according to Axelrod's memory of it, "I'll go first."

"Detoxication of Drugs and Other Foreign Compounds by Liver Microsomes," by Bernard B. Brodie, Julius Axelrod, Jack R. Cooper, Leo Gaudette, Bert N. La Du, Choco Mitoma, and Sidney Udenfriend, appeared in *Science*—then, as now, America's preeminent scientific journal, carrying papers read by scientists in every field—in April 1955. Studies showing that SKF 525-A inhibits the metabolism of a variety of drugs, it led off, had "suggested that the tissue catalysts responsible for their metabolism possess certain factors in common." Here was such a factor—the microsomal enzymes. Only as but one of several kindred findings did the paper refer to Axelrod's initial discovery.

Several years later, a review article in *Annual Review of Biochemistry* by Brodie and two colleagues acknowledged Axelrod's contribution: "The discovery of a microsomal enzyme system that deaminates amines stems from the observation of Axelrod that rats and dogs convert amphetamine to p-hydroxy amphetamines but rabbit liver microsomes yield phenylacetone and ammonia. Historically this was the first of a series of oxidative microsomal systems shown to require both oxygen and TPNH."

But that statement, insists Axelrod, appeared only at the urging of one of Brodie's coauthors, Bert La Du; "Brodie," he claims, "had left it deliberately vague." Besides, by then it was too late: To the scientific world, Brodie's lab had done it again, and Axelrod was just another faceless member of the scientific team fashioned and guided by Brodie's genius.

(Indeed, a 1981 review of the field—which, since those first papers, has continued to be an area of active research interest—credits "Brodie's Laboratory at the National In stitutes of Health" with having "initiated biochemical studies on the hepatic enzymes responsible for the oxidative conversion of lipid soluble compounds. . . ." Axelrod's name nowhere appears.)

It was a problem which had fascinated Axelrod and which he'd pondered constantly. To this day, he feels it was "the best work I've ever done," surpassing that for which he won the Nobel Prize. And here, he felt, he'd been cheated out of his discovery. He was incensed. "For years, he couldn't even mention Brodie's name," says Jack Cooper. Over the years he mellowed. But as recently as 1982, the issue was still sufficiently alive for him to write an account of the discovery in *Trends in Pharmacological Sciences* entitled "The Discovery of the Microsomal Drug-metabolizing Enzymes," setting out his version of events.

Priority disputes—battles over credit for discovery—have a long history in science. As Derek de Solla Price pointed out in *Little Science, Big Science*, had Beethoven never lived the great music of his age would have been someone else's, and quite different. Scientists, on the other hand, all seek solutions to the same mysteries, at least within a particular field. "There is," de Solla Price noted, "only one world to discover."

And whoever first reveals some aspect of that world walks away with all the marbles. Scientists do sometimes profess indifference to prizes, money, or fame, but almost never to the recognition of their peers. "You want to tell your story," one NIH scientist says. "You want to go to a scientific meeting, see your friends, and have them say, 'Oh, that's exciting. That's important'—and not find that someone else has already told the same story." Confirming the discovery of another counts for little; being indisputably first is all.

Was Julius Axelrod "first" to discover the microsomal enzymes? Was his simmering bitterness justified?

"You talk to ten successful scientists, and you could come up with a hundred examples of this kind of thing," says Park Shore, referring to the microsomal enzymes flap. Shore retired a few years ago from the University of Texas, where he'd gone after leaving NIH in 1961, and today lives in Santa Fe. At one point, as he sits out under the bright New Mexico sun, recalling his early NIH days, he pauses for a moment in thought, then abruptly gestures toward his home's just-completed patio: "Say you're designing a patio," he begins. "There's a group of you working on it, OK? You say, 'How about placing some bricks there?' 'That way,' you think to yourself, 'you could put a wall up right over there.' And then somebody else says, 'Yeah, and you could put up a wall right over there!'

"So who said it first? *Whose idea is it?* That's what happens all the time in science. When you have people working that closely it's impossible to say who thought of what." In the case of the microsomal enzymes, he says, "everybody was heading for it."

Sid Udenfriend, whose name was one of those to appear on the big *Science* article, agrees it was Axelrod who pointed the rest of them toward the microsomes. But then Brodie took it further, he says. Moreover, "it was an assigned project. It wasn't as if Axelrod decided to study microsomes when Brodie was away on vacation and that Brodie found a fait accompli. Axelrod did what Brodie asked him to do, although [compared to other technicians in the lab] he did it with more freedom." True, adds Udenfriend, "Brodie had never worked with an enzyme in his life. But then again, neither had Julie."

Jack Cooper, also among the seven names in the *Science* article, maintains that while Brodie did indeed treat Axelrod shabbily, Axelrod overreacted.

Axelrod is not alone among old Brodie hands to accuse Brodie of stealing glory rightfully theirs. But, says former second-in-command John Burns, the charge isn't warranted. In his natural enthusiasm, Brodie did have a way of moving in on something, he says. "At first, you welcomed it. Then you didn't even realize what was happening. In the end you sometimes felt excluded from what had initially been yours." But Brodie could always see in that initial modest idea or limited finding the germ of something grander. "He'd *make* it exciting. Sure, you'd hear grumbling: 'Hey, my idea's been taken over.'" And in a sense it had. But it had been made richer, too. Which is just the case, he says, with the microsomal enzymes: Axelrod came up with the initial finding, "but Brodie made it more exciting."

Whatever the merits of such arguments, Axelrod's anger at the time was unalloyed and white hot. This time, he vowed, he would get out from under Brodie for good.

6.
Separate Ways

IT IS AN AIRY, grandly scaled room with windows on three sides, on the sixth floor of a former library building at 2023 G Street in northwest Washington, D.C. It is called the Trustees' Room, and it is part of George Washington University. Here, on the evening of September 29, 1955, Julius Axelrod prepared to step before an eight-man examining committee and defend his qualifications for a doctor of philosophy degree. Among those on the committee was, as the formal announcement had it, "Bernard Beryl Brodie, Professorial Lecturer in Pharmacology."

The microsomal enzymes affair had pushed Axelrod over the line. He had to sever his ties to Brodie, had to become his own man as a scientist—and had to get his Ph.D.

But how, and where? At forty-two, he had little time to waste. One university he considered, Columbia, wanted him to take *three years* of coursework. That was out of the question. On the other hand, the catalog requirements of George Washington University (GW) seemed more flexible. GW was no Columbia, but it was certainly a respectable institution. And it was right there in Washington.

Axelrod talked to Paul K. Smith, GW's chairman of

pharmacology. "You've already got several Ph.D. theses," Smith told him, referring to Axelrod's two dozen or so published papers. So the thesis requirement, normally the biggest hurdle to a Ph.D., was no problem. And Axelrod already had a masters degree, so his course requirements would be few. He would, however, have to take a series of demanding comprehensive examinations in biochemistry, drug metabolism, pharmacology, and physiology. And, of course, satisfy the university's language requirements.

"The languages bothered me more than anything," Axelrod says. Not so much German, which he passed in the spring of 1954, but French. When the fearful day came, however, the journal article he was called on to translate was in a familiar scientific area, and he did fine. "I wondered why it had stopped me in the past."

During the 1954–1955 academic year, Axelrod was back in the classroom for the first time since 1941. One course was drug metabolism, which he could just as well have taught. In fact, he did teach parts of it. Today, he is fond of telling how in most of his classes, there'd be half a dozen medical students, some of them young enough to be his children, who did better than he. And how on one exam he was asked a multiple choice question about a drug called antipyrine, whose metabolism he'd studied at length and for which he had half a dozen papers to his credit. He got it wrong.

Beginning in late June of 1955, he took his comps, filling page after page of the spiral-bound exam booklets with detailed answers to deceptively brief one- and two-sentence questions like, "Describe the process of extraction as a method of separating compounds. Explain the principles and application of this technique to drug metabolism." He and the other four doctoral candidates were asked to survey antimalarial drug therapy; to compare paper chromatography with ion exchange resinography; to describe the latest pharmacological developments in the fields of mental health, motion sickness, cancer; to explain how a drug could be inactive in the test tube, yet active in the body. More than once—and necessarily so, if he wished to do the question justice—he cited his own research findings.

Occasionally, he'd indulge his fancy. Asked how one drug could be potentiated, or have its action prolonged, by another, he offered SKF 525-A as an example, noting it had been studied "by investigators working at the NIH, whose names I have forgotten." Asked to imagine the world of pharmacology in 1975, he predicted that "a homogenate of dried lizard skin, aged bat urine, and N-acetyl p-aminophenol will cure cancer of the stomach." Drug detoxification would be a solved problem by then, he foresaw, but "Axelrod will have relatively little to do with these developments, since he will be hard at work preparing for his ninth try at the pharmacology comprehensive examination. . . . These fantasies," he added, "are a result of the heat and fatigue."

All in all, his fears of grad school proved unfounded. He worked hard but, as he views it in retrospect, age and experience conferred on him an advantage: "I knew what was important and didn't waste time on the trivial," the way younger students often did. Compared to research, he found school almost relaxing, a pleasant break from the rigors of the lab.

The title of his hundred-page dissertation was "The Fate of Sympathomimetic Phenylisopropylamines." It looked like any other George Washington University dissertation, satisfying requirements as to typing margins, quality of paper, and the like. But as for its content, notes Herb Weissbach, "Julie just punched his reprints together." "It will be shown," Axelrod wrote in the introduction, that compounds including ephedrine and amphetamine "are metabolized to active pharmacological agents by a number of biochemical processes including hydroxylation, demethylation, deamination, and conjugation. . . ." It was the microsomal enzymes work cast into new form.

Axelrod felt no particular trepidation as he was admitted to the Trustees' Room that September evening; he figured he knew the material better than anyone. The only hitch came when he learned that smoking was prohibited; a chain smoker who has since quit, he almost turned right around and went home.

His inquisitors—he can't for certain place Brodie there—sat in comfortable arm chairs around a long table asking questions, while he sat at one end answering them. One, he recalls, was on how to chemically distinguish the levo and dextro forms of a compound, referring to substances that are mirror images of one another structurally but otherwise identical. "We were almost like peers," recalls Axelrod. He felt thoroughly relaxed. Later, he and two members of the examining committee went for a drink at a downtown D.C. bar.

That's all there was to it. He was Dr. Axelrod.

A Ph.D. in the sciences is normally a four- or five-year undertaking beyond the bachelor's degree, often more. Which explains why, years later, when Axelrod would tell scientific audiences, "I took a year off to get a Ph.D.," he'd invariably draw appreciative chuckles. What he usually didn't say was that it was a year off half-time. All the while he took classes at GW and prepared his thesis—and a decade before Timothy Leary began advising young people to tune into the cosmic hum on acid—Axelrod was studying the metabolism of LSD.

About when he started at GW, Axelrod had written the National Institute of Mental Health, which at the time was still formally part of NIH, and the National Cancer Institute, which still is, inquiring about positions. At NIMH his letter crossed the desk of Seymour Kety, the institute's new scientific director and organizer of its intramural research program. At the time, Kety was being deluged by hundreds of applications for his burgeoning program, mostly from young scientific unknowns. Axelrod's publications impressed him. He called Axelrod in to talk, came away "convinced that this was someone we should have."

But what would Brodie have to say? Brodie, says Kety, "had a reputation for being brilliant and for training people well, but also as rather dominating, and not one for giving his people independence." He was a powerful man, used to getting his way. Would he give Axelrod his blessing? Kety wasn't sure. But Brodie did come through with a generous recommendation. He said Axelrod was bright, capable, and productive and that, yes, he was ready to be independent.

Axelrod hoped to work for Giulio Cantoni, the discoverer of S-adenosylmethionine, a methyl-donating compound that had been found to figure in a variety of important life processes. But from all Kety knew of Axelrod's history with Brodie, he thought Cantoni an unwise choice; Brodie and Cantoni were too much alike. Instead, he promised to send Axelrod's application around to the various labs within NIMH. Sure enough, some time later Axelrod was approached by Edward V. Evarts, the young acting chief of the Laboratory of Psychosomatic Medicine, whose interest was in LSD-induced psychosis as a model for schizophrenia. Would Axelrod like to join his lab?

Axelrod was delighted. " 'But I have no background in schizophrenia or mental illness,' " he remembers adding.

"Don't worry," said Evarts. "You can do whatever you like."

•

You can do whatever you like. There was a whole philosophy embodied in that assurance, and it was Seymour Kety's own: NIMH was devoted to the relief of mental illness. But achieving that, its research director felt, required a far more intimate understanding of life processes than that yet achieved. Could you fix an automobile engine without knowing how it worked? Not usually. Similarly, he felt, "since the brain is so unknown, doing clinically 'relevant' research is futile." Therefore, he was prepared to let his investigators track down whatever research leads they wished. Faced with a choice between two equally promising ones, he was willing to bet, they'd choose the one with greater mental health relevance. The short-run payoff? Creative, motivated scientists would not be stymied. In the long run, basic science would gain, leading to clinical advances more abundant than if they'd been pursued directly.

Some years later, two clinical investigators, Julius Comroe and Robert Dripps, lent analytical force to Kety's intuition. The two undertook to examine the origins of the ten most important clinical advances in heart and lung medicine and surgery of the preceding thirty years. They tracked down 529 scientific articles that had, in retrospect, proven crucial to those clinical success stories. Of them, Comroe wrote, fully forty-one percent "reported work that, at the time it was done, had no relation whatever to the disease that it later helped to prevent, diagnose, treat, or alleviate." Penicillin, the anticoagulant heparin, and the class of drugs known as beta-blockers were among them.

While in 1955 the Comroe and Dripps findings still lay in the future, Kety's approach to mental health research plainly embraced them. It was an approach that risked leaving outsiders to science—like taxpayers, or Congress—nervous and mistrustful. Yet in the coming years it was to become the underlying research philosophy of NIH as a whole under the direction of James Shannon.

•

One Friday afternoon in the summer of 1955, Surgeon General Leonard A. Scheele called Shannon into his office and asked him to become NIH director. Fourteen years earlier, Shannon had been named research director at Goldwater. Nine years earlier, he'd appeared at a postwar news conference to announce development of new antimalarial drugs. Six years earlier, he'd moved down to the heart institute.

In 1952, having transformed the heart institute into a model scientific enterprise, Shannon had been named NIH associate director. In that capacity he'd dealt with the controversy surrounding the new Salk polio vaccine, just then being introduced on a national scale. A laboratory in California had manufactured a batch of defective vaccine. Several children had fallen ill. There were the makings of a full-scale scandal. "We had reporters sitting on our desks, using our telephones," Shannon's former aide, Bill Carrigan, remembers.

For Shannon, it had been a time of eighteen-hour days; he

found his weight dropping back to what it was when he ran cross-country at Holy Cross. But by all accounts he dealt with the problem masterfully. He closed down the vaccine manufacturers, reorganized NIH's role, instituted stiffer safety tests. In the pages of the *New York Times*, he predicted the problem wouldn't recur, and it didn't. Nationwide, the number of polio cases dropped from 28,000 in 1954 to 798 seven years later, and Shannon was counted as among those who, in the words of one account, "salvaged a wrecked vaccine program and set it on a true, safe course."

Now Shannon was being asked to take NIH's top job. He accepted.

"When can you start?" he was asked. "Monday," he said. And he did.

James Shannon, fifty, son of a Long Island farmer, had become, as one news report of his promotion put it, "commander-in-chief of the laboratory war on all important diseases: cancer, heart, arthritis, mental ills, etc." The Shannon Era was underway.

Shannon took command of a National Institutes of Health organized into individual research institutes. There were seven of them at the time—there are eleven now—each devoted to the study of one or another class of disease. Over the years their names have changed. The National Heart Institute, for instance, became first the National Heart and Lung Institute and then, in 1976, the National Heart, Lung, and Blood Institute. What was once the National Institute of Arthritis and Metabolic Diseases is today the National Institute of Arthritis, Diabetes, and Digestive and Kidney Diseases. This organizational twisting, turning, and expansion arose largely in response to what a *Washington Post* look into Shannon's NIH called "an unruly informal constituency of volunteer 'lay' organizations, each with its own political connection on Capitol Hill," and each lobbying for its particular disease. These were organizations like the American Cancer Society, the National Heart Association, and the Arthritis and Rheumatism Foundation.

Shannon's task was to align NIH's disease-oriented structure with the needs of basic research. The strategy he advanced all the years of his tenure as director was: Don't embark on a narrow search for disease cures at all. Instead, collect teams of bright, dedicated, curious researchers and set them loose on research problems of their own choosing. "Knowledge of life processes and of phenomena underlying health and disease is still grossly inadequate," he would write. Without such knowledge, it was a waste of time, money, and manpower to aim for the solution of a specific medical problem. He blamed the failure of polio vaccines back in the 1930s on lack of knowledge of the polio virus and techniques needed to culture it. He pulled the plug from an artificial heart program already approved because he didn't think cardiac functioning was well enough understood.

He didn't like the term *basic* research; he preferred calling it *fundamental*. But in the end it was about the same. As he put it in an article he coauthored for *Science* soon after becoming director, "The potential relevance of research to any disease category is [best] defined in terms of long-range possibilities and not in terms of work directed toward the quick solution of problems obviously and solely related to a given disease."

It was the approach to health research to which Comroe and Dripps would later offer convincing support. It was what Seymour Kety deeply believed. It was implicit in what Axelrod was hearing from Ed Evarts at about this same time. And it was to be instrumental in guiding the whole NIH research enterprise all through the Shannon Era and beyond.

•

By the time Shannon first came to NIH in 1949, a great pile of fresh mud had already accumulated on the hill behind Building 1, displacing goats and other experimental animals once housed there. This was the foundation for Building 10, the Clinical Center, where most of the scientific dramas of the Shannon Era were to be played out.

If Building 3 had the scale and feel of a college dormitory, Building 10 was like a big-city medical complex—a vast brick monolith, towering fourteen floors above the NIH campus, with as much floor area as twenty-eight Building 3s, crammed with cafeterias, libraries, offices, research labs, and medical wards, the whole complex criss-crossed by a labyrinth of intersecting corridors. There was nothing like it anywhere in the world.

When Shannon was trying to recruit James Wyngaarden from Harvard, he compared the Clinical Center, then in construction, to Massachusetts General Hospital's Ward 4, a ward reserved for patients of special research interest. The Clinical Center would be the same thing on a more massive scale—a "hospital" designed to serve the needs of research. Only Rockefeller Institute (now University) Hospital in Manhattan was anything like it, but it had only 40 beds, compared to 540 in the Clinical Center. The Pasteur Institute in France, the Medical Research Council in England—these preeminent national research institutions had no beds at all. "The NIH Clinical Center became for clinical investigation," Donald S. Frederickson would write, "what Gropius's Bauhaus in Dessau had once been for architecture."

At about the time the Clinical Center admitted its first patients, on July 6, 1953, most of the former inhabitants of Building 3 were moving into new, vastly larger labs in the new building. Taking possession of the seventh and eighth floors of the east wing was Steve Brodie and his Laboratory of Chemical Pharmacology.

•

It was a prodigiously scaled operation, the largest in the heart institute. Over much of the next decade, Brodie had six or seven section heads under him, each with four or five staff scientists, visiting scientists, or postdocs—perhaps forty researchers in all. It was so big that Brodie left the day-to-day running of it to a deputy—Burns, Costa, and finally James R. Gillette, who later succeeded Brodie as lab chief. His own office was in 7N 117, a room twice as long as it was wide, set back by a secretary's cubicle from the north corridor. Along the corridor, twenty-foot-deep labs branched to the right or left every two or three paces for a hundred feet. At the end of the hall, the Laboratory of Chemical Pharmacology marched right upstairs to the eighth floor, where the pattern was repeated.

LCP, as it was known, was immensely productive. "The papers came out like this," says Park Shore, his hands indicating a steadily mounting pile. Barbara Orlans, who joined LCP a little later, recalls how "we used to laugh at how prolific our lab was. . . . We counted up the B. B. Brodie publications for the year and reckoned that they averaged one every weekday." One every two weeks was actually more like it, at least if you restrict the list to those actually bearing Brodie's name. Still, with respectably productive scientists normally being good for two, maybe three papers a year, their output was astounding.

Beginning around 1955, the big stir at LCP was over serotonin. ("When the experiments were good, we called it sero*tonin*," Brodie would later recall, on receiving an honorary degree from the University of Cagliari in Italy. "When I heard it pronounced se*rot*onin, I knew the experiments were bad and I stayed home.") Serotonin is a substance in blood serum, long known to be able to contract blood vessels, that in 1953 several research teams had found in the brain as well. That same year, a University of Edinburgh scientist, J. H. Gaddum, showed serotonin's effects were blocked by LSD, whose chemical structure is similar. At a scientific meeting in London, he speculated that LSD produces its hallucinatory effects by blocking the action of serotonin in the brain, and that, as he put it, "serotonin might play an essential part in keeping us sane."

Park Shore was still a graduate student in Brodie's lab at the time, but it was he who actually got the idea. "Talk about 'taking a flier,' " he says, referring to Brodie's encouragement of experimental long shots, "this was an incredible flier." Shore had noted certain chemical similarities between serotonin and reserpine—which, beginning in the early 1950s, had come into wide use, along with chlorpromazine, for the treatment of schizophrenia. Shore injected some reserpine into a dog, collected its urine, and had Herb Weissbach, who was working with Sid Udenfriend on serotonin metabolism, analyse it. Weissbach came back with a urine sample fairly spilling over with a serotonin metabolite.

"He was dumbfounded," says Shore. Reserpine, the active ingredient in Indian snakeroot, a shrub indigenous to southeast Asia which for centuries had been used by Indian physicians to treat high blood pressure and mental disturbance, seemed able to free serotonin from its storage depots in the body.

That was the beginning. Then the big questions beckoned: Did serotonin release from the brain account for reserpine's pharmacological action? Was serotonin a central nervous system neurotransmitter, thereby playing a key role in brain function?

Around this time, Robert Bowman, another old Goldwater hand, had been trying to improve on the instrument Brodie and Udenfriend had first used during the war to measure Atabrine. Bowman's spectrophotofluorometer, as he called it, exploited the phenomenon of fluorescence, just as theirs had. But it could monitor a continuous range of ultraviolet light frequencies, not just a few fixed ones. And it was much more sensitive, down to one ten-millionth of a gram.

That sensitivity was needed now. Brodie, Shore, and their colleagues wanted to measure serotonin's concentration in the brain, and all existing methods were insufficiently sensitive. Bowman's shop-built prototype, which Axelrod was also using around the same time to measure LSD, was heaven-sent.

With it, they were able to conclude that reserpine achieved its antischizophrenic effects by releasing serotonin and that serotonin was probably a brain neurotransmitter. Soon Brodie was going much further, speculating wildly, advancing a theory that pictured serotonin and noradrenaline, another neurotransmitter, in a tugging and pulling match on different brain centers.

"Everything turned up gold for a long while," says Shore today. They were at a stage in the development of a new field where they could scarcely miss. "Things seemed very simple." Things weren't; some of their early conclusions were, as it turned out, only partially correct. And later evidence crippled Brodie's tugging-and-pulling theory. But wrong or not, it had generated enormous interest, broken the whole field wide open, and launched the new science of neuropharmacology.

Over the next few years, researchers came from Spain and Czechoslovakia, from Japan, Sweden, and even the Soviet Union to work in Brodie's lab. At one point, there were so many German-speakers that old Brodie hands refer to it as the "German Period." Sometimes, an eager scientist would fly in from Stockholm or Paris just to spend a week there. LCP, recalls Shore fondly, "was the Camelot of pharmacology. There were swarms of people. You practically had to fight them off."

•

"I lay down and sank in a kind of drunkenness which was not unpleasant and which was characterized by extreme activity of imagination. As I lay in a dazed condition with my eyes closed (I experienced daylight as disagreeably bright) there surged upon me an uninterrupted stream of fantastic images of extraordinary plasticity and vividness and accompanied by an intense, kaleidoscope-like play of colors." This fragment from an account of an acid trip comes not from the diary of a Haight-Ashbury hippie, vintage 1967, but from the lab notebook of Albert Hofmann, the Sandoz Laboratories researcher who in 1938 first synthesized Lysergic acid diethylamide, or LSD.

The move to NIMH had brought Axelrod down from Brodie's domain on the seventh floor to the third-floor lab of

Ed Evarts, who was using LSD as a kind of backdoor route to an understanding of mental illness. If, the thinking went, microgram doses of LSD could cause hallucinations and psychotic-like episodes, maybe LSD was mimicking biochemical processes involved in schizophrenia. Evarts had assured Axelrod that he had his pick of research projects. But LSD was a natural for the new recruit: It was a drug, wasn't it? And Axelrod knew all about drugs.

So he approached LSD as he'd learned to do from Brodie: First, find a way to measure the drug and its metabolites. Then, method in hand, track its metabolic fate. Piece of cake; he tackled it while still finishing up his doctorate. Mornings it was experiments. Afternoons, it was off to classes at George Washington. The first of two papers on LSD metabolism by Axelrod, Evarts, and two others appeared in the prestigious British journal *Nature* in 1956: "The development of a specific and sensitive method for the estimation of lysergic acid diethylamide in biological materials has enabled us to study its physiological disposition and metabolism. . . ."

By then, Axelrod had his doctorate and, as Ed Evarts says, "it had become plain that Julie ought to have a section of his own." In 1955, he became Chief, Section on Pharmacology, Laboratory of Clinical Science, National Institute of Mental Health. He moved into a lab on the second floor of the Clinical Center, room 2D45. "I felt pretty good about having my Ph.D. and a nice job," he says. He held the same job, remaining in the same lab space, with mostly the same drab, gray, government-issue steel furniture, until 1984.

At the beginning, it was quiet in 2D45. No one fought to come work with him, as they did with Brodie upstairs. It was just Axelrod alone, on his own at last, collaborating with one NIH colleague or another as the opportunity presented itself, but mostly sans technicians, postdoctoral fellows, or anyone else. He did drug metabolism studies. He worked on a class of compounds called glucoronides. He proposed a theory to explain narcotics tolerance. ("It stimulated a great deal of critical reaction," says Axelrod, "mostly negative.")

Then one day in 1956, at a departmental seminar, Seymour Kety told of a paper by a pair of Canadian psychiatrists named Hoffer and Osmond. It concerned adrenaline, the "fight or flight" hormone secreted by the adrenal gland. Adrenaline left out in the air turns pink, being oxidized to a compound called adrenochrome. Human subjects injected with this pink adrenochrome, said Hoffer and Osmond, hallucinate. They also claimed to have found adrenochrome in the blood of schizophrenics. Could it be that here, in the abnormal metabolism of adrenaline to adrenochrome, lay the biochemical basis of schizophrenia?

Along with manic-depressive psychosis, schizophrenia is one of the two broad categories of severe mental illness. A "primary disturbance of perceptual integration," is how one psychiatrist describes it. Bizarre behavior, withdrawal, hallucinations, delusions, and paranoia are its signposts. Yet it remains a diagnosis more slippery by far than, say, tuberculosis, with its satisfyingly specific bacillus as the known cause. It afflicts about one of every hundred people, and even with today's powerful drugs leaving most of them as outpatients, a quarter of all hospital beds are devoted to its treatment.

But only recently has schizophrenia come to be viewed as an illness at all. In the past, writes Robert A. Cohen, formerly of NIMH, "any serious mental disorder was popularly considered a reflection of a weakness, less respectable in kind and quite different in character from sarcoma, myocardial infarction, or multiple sclerosis." Devils, demons, and evil spirits have all borne the blame for it. Its "scientific" study has, in this century, largely been left to psychoanalysts and psychotherapists. But what of schizophrenia's biological basis? The question is itself of recent origin. The answer is, no one had a clue.

The abnormal metabolism of adrenaline. "This," says Axelrod, "struck me as a fascinating concept," and the ideal problem for him. For one thing, he'd previously worked on amphetamine and other drugs with structures related to adrenaline. For another, he was at the National Institute of Mental

Health, yet for the most part was continuing drug metabolism studies begun at the heart institute; he felt guilty about it. Studying the links between schizophrenia and adrenaline, on the other hand, would place him squarely within the area of mental health.

Axelrod figured that if the abnormal metabolism of adrenaline (to adrenochrome) was presumably to blame for schizophrenia, he ought to learn something of its *normal* metabolism. In the library up on the fifth floor he spent a day rummaging around for the details of adrenaline metabolism. There were no details. There were scarcely any general notions. The prevailing idea was that adrenaline was broken down by the enzyme monoamine oxidase, known to act in other drug-metabolizing processes. But it was just a guess.

The ignorance surrounding adrenaline surprised Axelrod. And it excited him, too: He was venturing into new, scientifically virgin territory—the world of the nervous system.

Adrenaline* is first cousin to noradrenaline, one of the two neurotransmitters of the autonomic nervous system: While lifting your arm or speaking are deliberate acts subject to conscious control, bodily processes necessary to life must take place without conscious intervention. The autonomic nervous system sees to it that they do, automatically regulating such functions as digestion, heart rate, and blood pressure.

Neurologists divide the autonomic system into two parts, the sympathetic nervous system and the parasympathetic, each of which, roughly speaking, opposes the other. The sympathetic system prepares the body for sudden bouts of muscular activity, as in sport, or battle, or chasing after a bus; it increases the heart rate, tenses muscles, expands the air passages of the lungs, enlarges the pupils of the eyes. The parasympathetic system has opposite effects, calming and quieting, becoming especially active during the digestion of food.

* Adrenaline is also known as epinephrine, and noradrenaline as norepinephrine.

Each system employs its own chemical messenger, or neurotransmitter. A neurotransmitter is the chemical that transmits nerve impulses from one nerve to the next. In the classic 1921 demonstration of neurotransmitter action, the Austrian Otto Loewi (who won the Nobel Prize in 1936 and later came to the New York University pharmacology department headed by George Wallace) placed two frog hearts in a common bath. When he stimulated the vagus nerve of one heart, slowing it, the beat of the second also slowed. Yet their only link was the common bath. Plainly, a chemical from the first heart had diffused through the bath and stimulated the second. Loewi termed it *vagusstoff*, later identified as acetylcholine, the "first" neurotransmitter—the neurotransmitter of the parasympathetic nervous system.

In 1948, Ulf von Euler showed that noradrenaline is the neurotransmitter of the sympathetic system. Noradrenaline is adrenaline minus a methyl group; the two are related, but not the same. Released locally, at the individual synapse, noradrenaline fires neurons one by one, achieving fine nervous system control. Adrenaline, on the other hand, is secreted by small glands that sit atop each of the two kidneys, the adrenals, and is dumped into the blood system as a whole. Acting at many of the same sites as noradrenaline, its effects are bodywide.

Now Axelrod was studying the metabolism of these sympathetic nervous system chemicals. For four months, he tried to confirm the Hoffer and Osmond hypothesis, looking for an enzyme that changed adrenaline into adrenochrome. He couldn't. His experiments were one disappointment after another. Then, one day in 1957, he opened up *Federation Proceedings*, the journal of the Federation of American Societies for Experimental Biology, to find a paper entitled "IDENTIFICATION OF A MAJOR URINARY METABOLITE OF NOREPINEPHRINE" by Marvin D. Armstrong and Armand McMillan. In it they reported that the urine of patients with a particular tumor of the adrenal gland contained large quantities of a compound called 3-methoxy-4-hydroxymandelic acid, which they named

VMA. And VMA, they said, was probably a metabolite of noradrenaline.

Axelrod seized on the brief, one-paragraph abstract. The "3-methoxy" part of VMA simply meant that attached to the third carbon of its benzene ring, replacing the usual hydrogen, was a methyl group linked in turn to oxygen. And it was this methyl group, a carbon with two attached hydrogens, that got him thinking: Take noradrenaline, lop off its amine, add a methyl, and what would you have? You'd have Armstrong and McMillan's VMA. And from where might the added methyl group come? Why, from Giulio Cantoni's S-adenosylmethionine, that's where!

S-adenosylmethionine, or SAM, had been discovered a few years earlier by the NIMH researcher with whom Axelrod had first asked Kety to let him work. SAM was apparently a universal methyl "donor," essential to a wide variety of life processes; whenever some step in a metabolic reaction needed a methyl group, more often than not it seemed to come from SAM, whose own methyl group was only loosely attached. Supply the right enzyme and it would cross over to another compound. Adrenaline, Axelrod suspected, was one compound to which it could stick.

But it was no sure thing. Because for adrenaline (or noradrenaline, by a similar sequence) to emerge several biochemical steps later as VMA, as Axelrod suspected it did, SAM had to affix its methyl group to a spot on the adrenaline molecule already occupied by a hydroxyl group. Maybe SAM could work that way. But if so, it had never been shown.

That very afternoon—it was March 10, 1957—Axelrod tried to find out. First off, he needed SAM. He had none. But he did have two compounds Giulio Cantoni had shown could combine to make SAM in the liver: the amino acid methionine and adenosine triphosphate, or ATP, the body's energy molecule. He combined this two-ingredient stew with a rat liver extract, then added noradrenaline. It disappeared.

He couldn't literally see it disappear. And yet, as he recorded the numbers representing it, he could almost feel it go: The original reading had been eighty. Now it stood at seven.

He repeated the experiment with various elements of the stew missing. When he eliminated the ATP, but kept the methionine, the noradrenaline remained. When he kept the ATP, but failed to add methionine, the noradrenaline remained. When he boiled the liver extract, rendering its enzymes inactive, the noradrenaline remained. Only with all the SAM-making ingredients present, and with the as yet unknown enzyme intact, was noradrenaline metabolized. "I knew I had it," says Axelrod. "I knew I had a new enzyme and a new metabolic pathway for noradrenaline."

He knew he had it. But still, he wanted to make sure. For one thing, he wanted to repeat the experiment with SAM itself, thus replacing an almost certain inference—that ATP and methionine had made SAM—with hard experimental fact. "Gabriel de la Haba in the laboratory next to mine generously gave me a little S-adenosylmethionine," is the way Axelrod has recorded the event. In fact, the story goes, Giulio Cantoni was jealously protective of his SAM, and Axelrod waited until he knew Cantoni was out of town before going next door to ask for some. He got it. It worked.

The noradrenaline was presumably being transformed into something that in turn changed to VMA. Axelrod knew what that intermediate something ought to look like, chemically, if he was right. So, simply assuming he was right, he asked Bernard Witkop and Siro Senoh, organic chemists in a neighboring lab, to make some from scratch. If noradrenaline broke down the way he figured it did, its intermediate metabolic byproduct, which Axelrod was calling normetanephrine, ought to be indistinguishable from what Witkop and Senoh were cooking up for him.

Three days later, Senoh brought him what he remembers as "beautiful crystals" of freshly synthesized normetanephrine. A standard laboratory technique known as paper chromatography—which leaves you with a large, specially prepared piece of paper full of spots, each representing a particular compound's tendency to migrate through various solvents—promised Axelrod a quick answer. If the spots were different, there was no point in going any further.

But the spots weren't different. They were the same. Which meant the compounds were the same. Which meant Axelrod's normetanephrine was noradrenaline's intermediate metabolite. Which meant that some hitherto unsuspected enzyme was responsible for the transformation.

Axelrod found the enzyme, isolated it, purified it. He discovered that it not only metabolized noradrenaline, but also adrenaline and dopamine, all three of which are members of a class of compounds known as the catecholamines. Axelrod named his new enzyme catechol-O-methyltransferase. (The *O* means that the methyl group it transferred was landing at a position on the benzene ring already occupied by oxygen.) Today, virtually any biochemistry, pharmacology, or physiology text will, among its diagrams of metabolic pathways, show "COMT" next to the little arrow between noradrenaline and normetanephrine.

Axelrod had barely started on his exploration of the nervous system. But, two years out of Brodie's lab, his maiden foray had led to the kind of textbook-altering result most scientists seek their entire lives.

"The best thing I ever did was hitching up with Brodie," Julius Axelrod would say years later. "The second-best thing was leaving him."

•

And the original Hoffer and Osmond theory? The paper that had sparked Axelrod's work, that had suggested metabolic differences between schizophrenics and normals? Seymour Kety had been skeptical from the first. "We'd been fooled many times before," he says, referring to too-simple theories about schizophrenia. "Besides, I didn't have confidence in their credentials as scientists. These guys claimed a faulty metabolism when we didn't know what normal metabolism was."

Still, he followed up on the Hoffer and Osmond paper by phoning a tiny firm named New England Nuclear, in Boston, and placing a ten-thousand-dollar order for a batch of tritiated noradrenaline. Tritiated noradrenaline is "hot" noradrenaline, noradrenaline made radioactive. Ordering it proved to be a momentous decision.

7.
Julie's Lab

THEY'RE EVERYWHERE. Most any biomedical research lab in the country is littered with them. Big ones on the doors of refrigerated rooms, small ones on instruments and glassware. One finds them on exhaust hoods, on storage cabinets, on waste cans—bright yellow stickers and signs, their familiar three-lobed design in red, reading CAUTION: RADIOACTIVE MATERIALS.

These radioactive materials don't, and cannot, fuel bombs or nuclear reactors; their radioactivity is too feeble for that. Instead, they serve as tracers in biomedical experiments: tag or label a compound with weak radioactivity, so that to the appropriate detector it may be said to glow, and you can trace its fate even in complex life processes. How intensely it glows tells how much of the original compound follows any particular biochemical pathway.

The tritiated adrenaline Seymour Kety ordered from New England Nuclear Corporation in 1957 was adrenaline tagged with a radioactive isotope of hydrogen called tritium. Many chemical elements are found in both stable and radioactive forms, or isotopes. The nucleus of the hydrogen atom, for example, normally contain a single proton, and no neutrons.

The rare tritium isotope is hydrogen with the same lone proton but *two* neutrons—three nuclear particles in all, the source of its name. Chemically, the two isotopes are almost indistinguishable, being active in the same reactions, forming the same compounds, and so on.

But tritium is radioactive, spewing out radiation that can be picked up by a liquid scintillation counter, a high-tech Geiger counter which brings each test tube into position, "measures" its radioactivity, records the value, then repeats the process with the next test tube, all automatically. Kety bought an early version of such an instrument at the time he placed his order for tritiated adrenaline and noradrenaline.

In 1958, the first batch arrived. Kety planned to use it to check for abnormal metabolites among schizophrenics, an idea prompted by the Hoffer and Osmond paper on adrenochrome that had launched Axelrod on his noradrenaline work. But now Axelrod had other ideas for the expensive radioactive compound just in from Boston. Why not inject a little into an animal, just to see where it went? "I confess I thought this was a half-assed idea," says Kety. What did Julie expect to learn?

Well, he did learn something. When he and two colleagues injected it into anesthetized cats, killed them, ground up their various body organs and ran test tubes containing the samples through the scintillation counter, they noticed that virtually none of the adrenaline wound up in the brain. It had reached the heart, and the spleen, and the pituitary, but not the brain. Conclusion? Adrenaline can't pass the blood-brain barrier—the capillary system that keeps some substances out of the brain and others in, and so serves as a biochemical cushion for the brain's mass of sensitive neural tissue. Since adrenaline was known to be in the brain, it was plainly being formed there from precursor compounds which *could* cross the blood-brain barrier.

A significant finding, and worth a brief paper; sent off to *Science* in December 1958, it appeared six months later. But the crucial data, the fabulous clue that would usher in a whole new understanding of sympathetic nervous system function, they saved for a later paper.

Axelrod had noticed it first with adrenaline and then again when he, visiting scientist Hans Weil-Malherbe, and Rockefeller Foundation fellow Gordon Whitby did a similar experiment with radioactive noradrenaline: They injected it into anesthetized cats, after two minutes decapitated them, then measured the noradrenaline that had reached each kind of body tissue. The distribution among the tissues was wildly unequal. In the aorta they found 33 nanograms (billionths of a gram) of noradrenaline per gram of tissue, in the heart 229. The pancreas had 46, the adrenal gland 150. The kidney 48, the spleen 229. What did this mean?

Axelrod thought he knew. For some time, a dissonant piece of experimental evidence had nagged at Axelrod's composure. Earlier, recall, he'd worked out how noradrenaline was metabolized, showing that not just monoamine oxidase was involved, but a new enzyme, the one he'd dubbed catechol-O-methyltransferase. Presumably, if you blocked both enzymes, thereby preventing noradrenaline's breakdown, its pharmacological effects ought to persist indefinitely.

But then a Food and Drug Administration researcher, Richard Crout, had actually tried the experiment and that wasn't what happened. Even with both enzymes blocked, noradrenaline stopped working. How could that be?

Nerve transmission is both an electrical and chemical process. The nerve impulse travels along the long thin filament of nerve in the form of an electrical "spike" of about a tenth of a volt. But it can go only so far before it encounters an obstacle—a gap between one neuron and the next. This gap, perhaps one ten-millionth of an inch across, is called a synapse.

The electrical signal reaching the synapse releases the neurotransmitter, which diffuses across the narrow synaptic gap to the next neuron, there to mate with a receptor shaped to receive it. In a way something like how a key fits a lock, thus opening a door, neurotransmitter and receptor together launch a series of electrochemical events that culminate in the firing of the second neuron. And so the nerve impulse is passed, like a baton, from neuron to neuron.

But how does the same nerve fire a second time? And a third? Nerve fibers can conduct a hundred or more impulses per second only if the whole elaborate electrochemical machinery is reset, in a few thousandths of a second, after each firing. Which means that after its quick swim across the synaptic cleft, the neurotransmitter must be inactivated or destroyed or otherwise cleared away. Can you take a second shot with a rifle while the spent cartridge from the first shot is still in the chamber?

In the case of the autonomic nervous system's other neurotransmitter, acetylcholine, the clearing-away is done by acetylcholinesterase, an enzyme that simply metabolizes it into something else. Block that enzyme, and, sure enough, acetylcholine's effects linger. Was it not likely that noradrenaline was inactivated in much the same way?

It was likely and it was logical and it was what everyone supposed, only it wasn't so. Crout's evidence was clear: Even when both enzymes known to be involved in noradrenaline's metabolic breakdown were blocked, noradrenaline was still being taken out of commission. To Axelrod this was vexing indeed. If enzymes alone were not how the body cleared away noradrenaline, what was?

Now, in the latest results, he had a clue: noradrenaline, he and his colleagues had discovered, didn't distribute equally among the various tissues. Rather, it concentrated in the heart, spleen, salivary gland, and adrenal glands—the very organs known to be most richly endowed with sympathetic nerves. Inject noradrenaline and these organs sucked it right up.

They did another experiment. They waited two hours instead of two minutes before killing the experimental animals and checking the tissue distribution. They found that noradrenaline levels scarcely dropped in the interim. It was known that after two hours noradrenaline no longer exerts a physiological effect, which presumably meant that it was by then metabolized into inactive byproducts. Yet there it was—

the data couldn't be more positive—unmetabolized, safe, and untouched, just sitting in those sympathetic tissues.

Could it be, Axelrod wondered, that after having crossed the synaptic cleft, mated with the receptor on the other side, and successfully fired the next neuron, noradrenaline was not destroyed by enzymes at all, but rather was reabsorbed into the original neuron and stored in some physiologically inactive form for use later on?

Nothing like such a mechanism was known to exist elsewhere in the nervous system. And, if true, it meant that what had been shown for acetylcholine didn't apply to its sister neurotransmitter, noradrenaline—an unexpected asymmetry of nature. How could Axelrod prove or disprove what he was coming to call the reuptake phenomenon? He and his colleagues exchanged many ideas. Then, one day, he and visiting scientist Georg Hertting—"a tiny, little, self-effacing man from Vienna," someone once called him—hit upon a simple way to do it: They'd take a cat and pluck out its superior cervical ganglia.

A ganglion is a clump of nerves. The superior cervical ganglia serve as a neural way station to certain organs, notably the eye muscles and the salivary gland. Axelrod and Hertting removed the superior cervical ganglia from one side of the cat's body, but not the other. To allow time for the nerves in tissues served by the excised ganglia to degenerate, they waited a week, then intravenously injected tritiated noradrenaline. An hour later, they killed the cat and took radioactive noradrenaline readings.

The results were striking: The eye muscles on the side of the body fed by the intact ganglia held 45 nanograms of noradrenaline per gram of tissue; the corresponding figure on the other side, where the ganglia had been plucked out, was 3.2. Similar differences applied to the salivary glands and other sympathetic tissues. When they repeated the experiment, this time waiting not a week but just fifteen minutes after removal of the ganglia, they found no difference; since the nerves had not had a chance to die, they continued to take up the injected noradrenaline.

The ring was tightening on a conclusive answer, that noradrenaline was being taken out of action by reabsorption into the nerve endings. That conclusion, if valid, neatly resolved a longstanding surgical mystery: Why does removal of sympathetic nerves make the organs they serve supersensitive to noradrenaline? Now the reason seemed clear. With the reuptake system rendered nonfunctional, any noradrenaline brought to the site tends to remain there, a continual irritant, as it were, to the nerve.

In the experiments which followed, Axelrod further elaborated on his first, crudely pencilled-in sketch of the reuptake phenomenon. He and his colleagues stimulated nerves of the spleen and watched as previously injected "hot" noradrenaline was released into the bloodstream. They found that cocaine works by blocking noradrenaline uptake, thus allowing the neurotransmitter to exert its effects longer. They pinpointed the anatomical structures within the nerve endings in which noradrenaline is stored, tiny granulelike vesicles they could actually see with an electron microscope. They showed how antidepressant drugs work by making available to the brain more noradrenaline. In paper after paper over the next few years, Axelrod and a growing cast of collaborators broke open a new field.

Not that their work progressed as neatly and inevitably as all that. "I have notebooks full of experiments that were ambiguous and led nowhere," Axelrod says—tedious, endlessly frustrating weeks and months in the lab that succeeded not even in saying how nature *didn't* work.

Today, a basic description of sympathetic nerve function might occupy four or five concise, authoritative paragraphs, perhaps a page, in a physiology text. Seeing it there in black and white, bound between hard covers, its credibility enhanced by the power of print, it is easy to forget that once the subject was shrouded in profound ignorance. Ulf von Euler had shown that noradrenaline was the neurotransmitter of the sympathetic nervous system; beyond that, virtually nothing had been known. Axelrod treated this naturally occurring substance as if it were a drug, employing the same basic drug metabolism strategy he'd learned from Brodie: Find a way to measure it, then trace its fate.

"The Fate of H^3 [tritiated] Norepinephrine in Animals," by L. G. Whitby, J. Axelrod and H. Weil-Malherbe, appeared in Volume 132 of the *Journal of Pharmacology and Experimental Therapeutics* in 1961. It was thirteen years after "The Fate of Acetanilide in Man," by Brodie, B. B. and Axelrod, J., had appeared in the same journal.

•

Work progressed. Papers appeared. And, gradually, a change could be seen taking place in Axelrod's lab. First it had been Weil-Malherbe who'd worked out of 2D45 for a few months. Then the Rockefeller University fellow, George Whitby. Then Georg Hertting from the University of Vienna, and the first of the research associates, Lincoln Potter, with whom Axelrod discovered the noradrenaline storage sites. Later, Whitby, back at Cambridge, asked his student, Leslie Iversen, to work on a problem he'd begun with Axelrod; Iversen came to Bethesda. And Axelrod heard about a Frenchman, Jacques Glowinski, who'd developed a method for injecting noradrenaline into the brain; soon Glowinski was working out of 2D45, too. And more research associates began coming through. Julius Axelrod, for so long the student, the assistant, the apprentice, now was mentor to others.

From the late 1950s and early 1960s on, it became fair to say there was such a thing as "Julius Axelrod's Lab." Oh, it was nothing like Brodie's vast domain upstairs. Axelrod was just a section head, and would remain so, repeatedly declining promotions that would take him away from the lab bench. "To stay small was a way for him to stay productive as a scientist," is how Michael Brownstein, who joined Axelrod much later, explains his mentor's attitude. "It meant a lot to him to go into the lab and work on his own ideas." Up until he won the Nobel Prize, Axelrod was at the bench himself every day. He had no legions of obedient underlings to do his bidding, no fixed team of subordinates. Still, as someone would say years later, "in that tiny, cluttered, crowded lab he hatched the revolution in the neurosciences."

For years after Axelrod went out on his own, many at NIH and elsewhere still thought of him as Brodie's old technician. Jack Orloff was one of them. "I assumed he was a super-technician, someone who knew a little about biology and a lot about chemistry," he says. But soon Axelrod emerged as a respected scientist in his own right. Victor Cohn, a member of Brodie's lab during the late 1950s and now a pharmacology professor at George Washington University, recalls how while Julie himself had little to do with Brodie, a contingent from Brodie's lab would visit him regularly to talk about drug metabolism.

By all accounts, Brodie felt keenly competitive with his former technician (as he did with Sidney Udenfriend, also emerging as an important scientific figure during these years). When Axelrod came out with a new paper, Brodie pounced on it, reading it with special interest. And he discouraged a free exchange of ideas between the seventh floor and the second. "It was difficult to talk freely," Mimo Costa remembers. When Costa first joined Brodie's lab in 1960, Brodie asked him to organize a seminar. Innocently, he invited Axelrod. Don't make that mistake again, he remembers Brodie telling him.

Meanwhile, young scientists were coming to work with Axelrod, then leaving a couple of years later to spread the word about "this funny little gnome of a man," as someone once called him, whose loose, seat-of-the-pants approach to doing science was not only immensely productive, but grand, good fun. Jack Orloff remembers traveling to Europe during the 1960s and coming away "surprised to know in what high regard he was held in Europe. There, Julius Axelrod was *the* guy."

•

It is more than a quarter-century since John Daly first met Axelrod. "All the biochemistry I know I blame on him," he says, smiling. Then, more seriously: "I owe him so much."

Daly's tiny, ceramic-tiled lab in NIH's Building 4 uses space the way a skyscraper does: vertically. Canyons of papers and books rise up on every side of the scant stretch of open floor in the middle, a cascade of centrifuges and boxes and journals and molecular models that threatens to push up through the ceiling. The daylight barely peeks through venetian blinds half-obscured by black metal shelving heaped with reagent bottles. At one short, clutter-free length of bench, across from a large fluorescent-lit terrarium full of brightly colored frogs, sits Daly, thin and bearded, looking like an overage graduate student in his striped oxford shirt with tan leather workboots, recalling his early days with Axelrod in the late 1950s and early 1960s.

At first he felt like an outsider in Axelrod's lab; Daly wasn't Axelrod's student at all but had originally come to NIH to work with organic chemist Bernard Witkop, whose lab had synthesized the crystals of metanephrine Axelrod had needed for his COMT work. Daly became involved and soon was showing up at Axelrod's lab two or three mornings a week. He was twenty-five, had just earned a Ph.D. in chemistry from Stanford, but knew scarcely any biochemistry. Axelrod taught him.

"There was Axelrod with his lab coat on, personally showing me, pipetting things himself, doing the techniques one-on-one. It was more gratifying that way. With a lot of people, you come to a lab and there's a big name, and you're lucky if you can see him once a week or once a month." That's how it had been with Witkop, whom he went two months without seeing. Axelrod's lab, on the other hand, was small and informal. You didn't have to set up a time to talk with the Chief; he was right beside you, at the next bench, always interested, always enthusiastic.

Even writing a paper was different with Julie. In many labs, you'd write a draft, then submit it to the lab chief, who'd tell you what he wanted changed. Then you'd go back and revise it. But with Axelrod, you actually sat down and wrote it together, right there, at his gray metal desk under the window. Daly couldn't get over Julie's willingness to put in so much time with him.

For three years, he migrated regularly over to 2D45; for another three he and Axelrod collaborated intermittently. In the end they published eight papers together, mostly on methylation pathways. "Those were the golden years," recalls Daly, the years that saw the initial influx of young men who would one day style themselves Axelrod's scientific "children" and who, over the next two decades, would spread across the country and around the world.

"They were a tremendous group," says Axelrod of that first group of research associates and visiting scientists. "They all became famous."

The first of the research associates was Lincoln Potter. A recent graduate of Yale medical school, Potter had just finished a brutal internship at Brigham and Women's Hospital in Boston, where he was used to getting four hours of sleep a night. Now, as a research associate, part of an NIH program designed to seduce promising young M.D.s into research, he was an officer in the Public Health Service. His time was his own. He made a decent living, could afford a little house for his wife and child. In Bethesda, unlike gray, dour Boston, the sun seemed always to be shining. For him and the other research associates passing through the lab, "it was an illustrious and delightful time of our lives."

Potter remembers 2D45 as a tiny lab with Julie in one corner and a test tube rocker, for mixing ingredients, forever clattering away. "Uh, this is what's happening," Axelrod told him by way of introduction, and pretty soon Georg Hertting was handing him a syringe and there he was, an arm's length from Julie's desk, injecting radioactive noradrenaline into rats. "Julie was on top of the results every minute, right there in the lab with us. As Georg and I worked on the norepinephrine, Julie worked on his methylating enzymes."

The first thing you'd see as you walked into Julius Axelrod's lab was Julius Axelrod at his desk, directly across a clear expanse of tiled floor from the door, slouched in his government-issue swivel chair, chin resting on chest, his glasses pushed up onto his forehead, a journal article pulled up close to his single functioning eye. He seemed somehow vulnerable like that, insulated by no secretaries, set off by no walls, right in the middle of everything. And that's the way it was for almost thirty years.

"The only thing you're going to see of him is a locked door saying, JULIUS AXELROD, NOBEL PRIZE WINNER," Juan Saavedra remembers a friend warning him just before he joined Axelrod's lab early in 1971. Well, says Saavedra," "not only did he not have a locked door, he didn't even have an office. He was not Professor Axelrod, but Julie. And not only could I see him, I ate lunch with him every day. And I could ask any question I wanted, even stupid ones."

As his younger colleagues mixed chemicals, set up experiments, and recorded data a few feet away, Axelrod went about his business, apparently undistracted. ("I owe it," he says, "to the New York subways," where he used to study on the way to school.) After winning the Nobel Prize, he could have had any kind of office he wished. He opted for none. What was an office, after all, but a place in which to write, away from the bustle of the lab? And he liked the bustle and traffic of the lab, the frequent interruptions. "If they interrupt," he reasoned, "they must have something important to tell me."

Something important to tell Julie. That's what his students lived for, competing with one another to supply it.

When Axelrod got excited about something it was like the sky had lit up. His enthusiasm for intriguing data and his encouragement of those furnishing it were legendary. Axelrod's special gift, a student once said, was that he could always convince you that whatever you were doing was earth-shakingly important. Another veteran of the lab, now himself a lab director, observes that a young scientist needs, most of all, "somebody to tell you you're good. Encouragement is very reinforcing, very important in training young people." In Axelrod's lab he got gobs of it.

I saw how myself, once, when Axelrod was explaining the design of a key experiment. Sometimes I'd been able to easily follow his accounts of his research, sometimes only with difficulty. This time, I was just barely hanging on, barely following, when suddenly the point he was leading up to clicked into focus, whereupon I completed his unfinished sentence. "Exaaactly!" he exclaimed, his face aglow. I knew that his delight didn't reflect on me, personally, but rather on that instant of triumphantly shared communication. Yet it felt otherwise, as if I must be a very special person to please him so.

There was "an informal hierarchy" around the lab, says Michael Brownstein, who later went on to head his own lab at NIMH, your place in it varying from day to day depending on how Julie viewed your work. If he wandered by two or three times a day, your status rose; you knew he was interested. He'd go over the data with you, get excited, and soon be telling you all the great experiments you could be doing two months from then, firing off one idea after another.

Sometimes, he could get on your nerves that way. "Julie can be kind of a *noodge*," says Brownstein, using the Yiddish word for nag or pest. You might scarcely have set up your experiment and there he was, all hepped up, breathing down your neck for the data.

Far worse, though, was when he wasn't interested at all. His encouragement carried weight, needless to say, only because it was not dispensed lightly. "If you hear enough, 'Oh, this is really interesting,' but there's no content to it, it has no value as coin anymore," says Brownstein. But Axelrod dollars, as it were, never suffered from inflation. When you had little to show for your efforts, you could see it in Julie's face. It was like the sun had shut down. "There was usually," as recent sometime-collaborator Merrily Poth says, "a fair-haired boy" in the lab, someone on whom Julie's light fell with special warmth. "Oh, have you seen his data?" Julie would come to you, all excited. Except that you wanted to be the one whose work he trumpeted. "It made you want to run back to the lab and work harder."

Axelrod felt uncomfortable when he first took on his own students; he'd been in their boots too long himself. Hans Thoenen, a tall German of Lincolnesque gauntness who worked with him in the late 1960s, credits Axelrod with "profound tolerance. He took us with our weak points, considered us as established, competent, and equal partners"—no doubt, suggests Michael Brownstein, because "Julie realized the damage that the massa-boy relationship can have."

Axelrod tried to give his students problems at which they were apt to do well, yet not so transparent as to be trivial or dull—a tricky balance to achieve. Your first project, says Brownstein, was "something at which you might succeed in a month or two, where you'd be able to land on your feet, running, and get lots of strokes from Julie. In the meantime you could think about what to do next.

"That strategy worked," Brownstein adds, "for as many people as he tried it on."

Axelrod's young colleagues were often just out of medical school, where they were used to grimly bearing down, absorbing great gobs of new material, bleary-eyed and miserable much of the time. Most needed convincing, Axelrod says, that there were no exams, that they could just relax and enjoy what they were doing. This wasn't school, where all the right answers were known and one had only to digest them. Rather, they were exploring uncharted territory, with few signposts to go by. Research wasn't a grind—or at least it didn't have to be—but an adventure into the unknown.

Axelrod showed them, says Jacques de Champlain, now at Université of Montréal, that science "can be a creative act, discovery a source of joy." Says Lincoln Potter, "It was that sense of wonder, magic, discovery, and delight that we had when we were kids that Julie brings to science."

"Just follow your nose," Axelrod would say. Research, as he approached it, was no grand, elaborately structured affair where you thought out everything beforehand and did everything with meticulous care. No, it was trying this, trying that, going where the muses moved you, guided as much by intuition as by logic.

Faced with choosing between a fresh, important problem and a minor one that's been picked to death, many scientists will go for the stale, the trivial, and the small. Perhaps because it's safer, the route better marked, the results more certain. Axelrod, with what to his peers seemed unerring instinct, picked the good problems, the meaty ones, and picked them when they were still largely untouched by others. "You have to ask the important question at the right time," he says. "Ask a year later and it's obvious. You've got to ask before it becomes obvious."

Axelrod, explains Michael Brownstein, would focus on the most robust phenomenon, one easy to study, and important, and leave to others that were neither. "His idea was, you paint with broad strokes. Then, if someone thinks it needs more detail, let him fill it in himself." He would, in other words, scientifically "skim the cream," as generations of Axelrod students would pass along his precept.

All you needed were good ideas and the willingness to try them. What you didn't need, and didn't want, was too exhaustive a knowledge of the existing literature—because, as one former student puts it, "all it can do is tell you what you can't do." As in so many areas of his scientific style, it was as if Steve Brodie were speaking across the generations, or at least down from the seventh floor of Building 10.

"You don't learn anything by thinking about what to do," Axelrod would say, "just by going into the lab and doing them." The experiment doesn't support your idea? Too bad. Try something else. You always have to be ready to drop a cherished theory, no matter how long you've worked at it. You can't get too emotionally involved. "You have to give up in the face of the facts." He was infinitely adaptable, at home with ambiguity. Research, to him, was one crime of opportunity after another. He was, as it were, "unprincipled," rudderless, willing to go wherever the wind took him.

"If you can get your foot in the door, you know where to go," Axelrod would say. "You just follow your nose, and go from one thing to the next, one step at a time," never getting stuck or slowed on the unmanageable and the difficult. If he could measure it, he'd do it, says Donald Brown, who worked across the hall from him in Building 10 and occasionally collaborated with him. That was his foot in the door. Method in hand, he'd ask, "How much is in the brain, how much in the liver?" And how rigorously did he work in getting answers? "Rigorously enough," says Brown, smiling. Not for Axelrod the biochemical tradition of purifying everything down to the last molecule.

Axelrod never proved anything, not really, but rather was content to confirm an inference by various approaches and let it go at that. Which usually meant going onto the next problem without being absolutely sure of what he had. For those not so confident of their scientific intuition, nor so ready to rush ahead to the next step, it could be unnerving, like stepping along an icy sidewalk, the pavement forever threatening to slip out from beneath your feet. Scary? "Damn right," says Martin Zatz, who first joined Axelrod's lab as a research associate in the mid-1970s and now works in Michael Brownstein's Laboratory of Cell Biology at NIMH. "Can I work the way Julie works?" he'd sometimes wonder. "And I'll think, 'I want to, but I can't.' "

"Julie plays with ideas the way a kid plays with toys," says Brownstein. "He doesn't bowl you over with intelligence; he never would be first in his class at Bronx High School of Science. But he has a gift for following up important things." His peculiar specialty was, as Brownstein says, "quick and dirty experiments based on his own special insights." And there lay the danger for anyone wishing to ape his scientific style—"that your insights won't be as good as his, that you'll miss things."

What you could never teach in a book and what his students were there to absorb were Axelrod's scientific instincts. Irv Kopin, now a distinguished NIMH researcher in his own right, began working with Axelrod in the late 1950s. "I remember that little blackboard behind his desk," he says. "That was where most of the basic breakthroughs in the neurosciences were worked out first. . . . Julie would pull things out of the sky and you didn't know from where. He always found it hard to explain, too. He'd say, 'Follow your nose.' *But you had to have the nose.* He had it. He had it about people and he had it about scientific problems. He knew when a problem was too difficult to solve and when it was ripe. But he'd never say this is too difficult *because* of A, B, C, D. It was more intuition."

Even at the lab bench he "had it." Sometimes he used the same pipette all day long, a sloppy practice that invites contamination. But he knew that a trace of contamination wouldn't mask the kind of results he'd set up the experiment to reveal with black-on-white crispness. By every account, he was a master at reducing complex questions to simple experiments with clear answers.

Roland D. Ciarnello, now at Stanford, tells how he can never see the *New York Times* without thinking of his mentor. They'd sit at Axelrod's desk going over the data and the moment Ciarnello got them bogged down in useless detail, Axelrod's attention would drift. "When you got to the differential equations, his eyes would wander to the *New York Times* on his desk. At that point I knew I had to simplify further."

"I don't like to do complex experiments. I'm not a complicated person," says Axelrod, as evenly as you could imagine, mildly, as a statement of neutral fact. But another time, extolling the virtues of simplicity in science, some of the modesty falls away. "Picasso," he says, "makes a single line—but it takes a lot of time and thought."

Axelrod's scientific articles were sometimes almost laughably simple. A few numbers, a few bar graphs that looked like they'd leapt intact from a sixth-grade arithmetic book. Axelrod had no use for statistics; resort to them, he felt, meant simply that the experiment was poorly designed. Better, results so plain they fairly shout out their truth: noradrenaline on the side of the body fed by intact cervical ganglia, 45; on the other side, where they'd been plucked out, 3.2.

Likewise his presentations at scientific meetings: no bludgeoning of his audience, just the essentials. Scientists who throw every last bit of evidence into grant proposals, the better to counter possible arguments by skeptical reviewers, often do the same for oral presentations, Axelrod has noticed. Not necessary, he's found. Omit the tedious details. Keep it simple. "They'll believe you," he says.

In the early days especially, Ed Evarts reports, some looked down on Axelrod because he did not seem intellectually sophisticated. He preferred the simple to the complex, was never absorbed or sidetracked by details. "He'd go step by step, always doing things that gave him feedback." But it was never "step by step" like lemmings in grim lockstep marching to the sea. Rather, it was step by tentative step through a minefield of ignorance, probing here and there for an opening before scurrying ahead.

The alternative research strategy, says Evarts, is to carefully design a project in advance, setting out a sequence of experiments which, slavishly followed, presumably guarantee a grand conclusion at the end. Not Axelrod. He relied on intuition rather than brute logic and elaborate planning, never locked himself into experiments that had to be done regardless of how earlier ones came out.

His experimental strategy was loose, flexible. "He put it together every afternoon."

•

And every afternoon—at five, or four, or sometimes on Fridays even earlier—he left for the day. "I wasn't a hard worker," he likes to say. At home, he'd play with his two boys, read—books, magazines, newspapers, scientific journals—and "sometimes," he adds, "think about my problem."

He tosses it off lightly, but a talk he gave in 1971 on the occasion of the 150th anniversary of his alma mater, George Washington University, hints that "sometimes" may understate the matter. Essential to research success, he said then, is neither outstanding scholarship, nor exceptional intelligence, but rather motivation and commitment. That "does not necessarily mean working in the laboratory day and night, but you think about the problems you are currently working with all the time, no matter what other activity you are engaged in. My wife occasionally complains that I give an inappropriate response to her question because my mind is elsewhere. I might add that some of the best ideas come not in the laboratory but as I am trying to go to sleep, listening to boring lectures, or while shaving."

Axelrod's comment represented what was for him a rare mingling of the personal and the professional. Unlike Brodie, he kept those two realms scrupulously apart. Current NIH director James Wyngaarden remembers how, even back in their car pool days, Axelrod had an arrangement with his wife to that effect. Saturday mornings belonged to him. But come noon, he'd close his books, put away his papers, and metamorphose into a family man for the rest of the weekend; Wyngaarden would sometimes see him shopping or doing laundry on Saturday afternoons. The arrangement, he remembers being aware, was one Axelrod had specifically "negotiated" with his wife, Sally.

Sally Taub Axelrod is, by all accounts, a fiercely private woman of egalitarian, even ascetic bent, genuinely offended by pomp and spectacle. She clings resolutely to her privacy —declining, for example, to be interviewed for this book.

Of Hungarian Jewish stock, she came out of the same Lower East Side streets as Julie, was imbued with the same thirst for intellect, the same progressive social values. And she's retained it all her life, says Axelrod, who, introduced by a mutual friend, married her in 1938. A graduate of New York's Hunter College and a teacher by training, she's taught second grade, worked with illiterates and with the retarded. "A good woman, a committed sort of person," is how her husband pictures her, one offended by hypocrisy and inclined to see things in black and white.

The day of the Nobel Prize announcement in 1970, she was at a teacher's convention in Baltimore. "MRS. AXELROD—URGENT!" blared the public address system. And all the way back to Washington she fretted over the public spectacle, the intrusive phone calls, the disruption to her personal life the Nobel was sure to mean. For all the fuss and bother, she was later rumored to have felt, she would sooner have seen Steve Brodie get it.

But a few days later, at the NIH recognition ceremony, Axelrod expressed thanks to his wife, "who bore up with me when the experiments didn't pan out"; Sally was right there in the first row to hear him. During the painful microsomal enzymes episode, he reports, she "heard about it every night," down to the technical details. She encouraged him to pursue his Ph.D., then later drew the job of keeping the kids away while he studied for it. At the Nobel ceremonies in Stockholm, warring against her temperament, "she handled herself," as Axelrod puts it, "with great style."

Sally Axelrod is said to have sometimes felt pangs of remorse for having not, in her own eyes, more graciously received her husband's research associates over the years. Indeed, even those of them closest to Axelrod say they never so much as set foot in his apartment. What they learned from him they learned in the lab, not over drinks after work or in casual dinner table patter. "Julie's conversations with you focus on data, mainly," says Michael Brownstein. The personal dimension? "He took about as little interest in that as anybody I've ever known." Likewise, he preferred his own personal life to remain off limits to others.

Still, he is almost universally revered by his students, many saying they made the best decisions of their lives in joining him. If the dominant feeling for Brodie around the lab was respect, that for Julie was love.

"Julie stories" are boundless in number and invariably affectionate. One time, Roland Ciarnello recalls, the two of them were at lunch in the Building 10 cafeteria downstairs, reviewing the results of an experiment Axelrod had suggested. "Julie," he said excitedly, "this confirms your theory that—" "Einstein had a theory," Axelrod interrupted. "The rest of us do the best we can."

He was gentle, he was soft-spoken, and he never got angry. Ed Evarts tells of the time when he, Seymour Kety, and Axelrod were in Colorado for a scientific meeting. Just back from a drive in the mountains, which left Axelrod feeling queasy, they were getting out of the car when the door slammed shut on Axelrod's hand, making a painful mess of his finger. "But all he did was look around a bit quizzically," says Evarts. "He didn't even say 'Jesus Christ,' or 'son of a bitch.' I guess he was just glad the ride was over."

Years later, on the occasion of a testimonial dinner for Axelrod, Jacques Glowinski, with whom Axelrod first extended his study of the nervous system to the brain, came all the way from France to pay homage to his mentor. At the head table in the Chevy Chase Women's Club that evening were Seymour Kety and Irv Kopin from NIMH, Congressman Steny Hoyer, Senator Thomas Eagleton, George Keyworth, President Reagan's science advisor, and many more, each of whom got up to say a few words. NIMH scientific director Fred Goodwin, presiding over the affair, had allotted each of them two minutes, a guideline until then observed. But Glowinski was not about to be so obliging.

"It is difficult to say in two minutes what I have been feeling for twenty years," he said with an insouciant nod to Goodwin. "So I will take my time." What he proposed to do, he said, the hall hushed at the Gallic chutzpah of Glowinski's podium coup, was to "write," then and there, a book "of the rules of how to do research"—Julie-style.

Chapter One, he began, his command of English idiom a little rusty, was "How To Do the Lab." "First," he said, "you must have a very small lab. You must have a desk, a very old desk. And then, a balance near the desk. That's essential. You must have glassware, but not a lot. You need SAM [S-adenosylmethionine, of course]. Without it, you never will succeed.

"You don't need equipment. It breaks. Better to get it from other labs. And you don't need many technicians. Do it yourself. Make it simple."

Subsequent chapters offered advice, likewise à la Julie, on working hours, recruitment of junior colleagues, research strategies. Quoting Axelrod, Glowinski advised "never to write the paper before doing the experiment." And, "Don't cry if your paper is rejected [as Axelrod's sometimes were before he won the Nobel], just because it is original."

In a fifth and final chapter, Glowinski offered an Axelrod maxim well known to those who'd worked beside him: "For a scientist, what is better, a good experiment or . . ." Giggles erupted from the audience as he made a show of groping for language polite enough to represent the original. Finally, Glowinski settled for, ". . . or a good love affair?"

His book offered one last bit of advice: "You have not to run after the Nobel Prize," he said, "only to wait for it.

"And," he added, "go to the dentist."

•

Axelrod won the Nobel Prize in December 1970.

At about the same time, forty miles up the expressway in Baltimore, a young graduate student, Candace Pert, had just joined the laboratory of a Johns Hopkins University pharmacology researcher named Solomon Snyder.

Snyder, at thirty-one the youngest full professor in Hopkins history, was unlike any scientist she'd known before. His lab was charged with a heady, devil-may-care exuberance. With him, experiments somehow seemed riskier, results less predictable, mistakes more common; but when those three bars did line up in the slot machine window, then, by god, you'd really hit the jackpot. Science, once to her "so serious and technical," dressed in a dour gray, now exploded with sparks of orange, green, and gold.

Snyder, she noticed, never himself worked at the bench. Rather, he preferred the role of idea man—proposing alternative research strategies, suggesting techniques, seeing hidden implications in stray pieces of data, giving each small part of a project its place in the Big Picture. He gave his students independence, yet managed to keep them gently in rein. He'd prowl the lab, making the rounds of his various grad students and postdocs, hoisting himself onto a lab bench, questioning, suggesting, and then sometimes, on a tricky problem, wondering out loud, "What would Julie do in this situation?"

"Julie," of course, was Julius Axelrod. Sol Snyder's way of doing science was Axelrod's way, just as Axelrod's was Steve Brodie's.

Snyder had been a research associate with Axelrod between 1963 and 1965, emerging as his most devoted disciple, the most faithful to his scientific style. Snyder was no clone of Axelrod; their personalities were very different. Nor had he himself ever worked in Brodie's lab, though he did meet him a few times. And yet, he emerged from his two years in 2D45 with some part of the scientific legacy Axelrod had received from Brodie passed down to him. Deflected from a safe, predictable career in psychiatry, he found himself propelled into a world of neuroscientific research where the stakes were high, the competition keen, and the adrenaline freely flowing.

During this period, Axelrod and Snyder were down on the second floor of Building 10. Brodie was up on the seventh floor, director of the Laboratory of Chemical Pharmacology. Shannon, the man who'd given Brodie his start, was over in Building 1, boss over them all. For those few years in the early 1960s, the mentor chain lay intact, all its links together.

8. The Golden Era

ONE DAY IN 1957, Donald Brown walked into a guitar shop at Eighteenth and M streets in downtown Washington, D.C., to inquire about lessons. Behind the counter stood eighteen-year-old Solomon Snyder, who quoted the rates charged by the shop owner. They were higher than Brown cared to pay. "Look, I'll do it for you for much less," said Snyder. Better than that, he'd come round to Brown's apartment in Bethesda to give them. "Why not?" thought Snyder's new pupil.

As a teacher, says Brown, young Snyder was "very patient. He really had hopes he could teach me something." But Brown, at least in his own estimation, was hopeless, and after a while the lessons degenerated into as much talk as music-making. Brown, twenty-six, was an M.D. from the University of Chicago who'd come to Washington for two years as a National Institute of Mental Health research associate. Snyder was a student at Georgetown University. Soon, the two weren't talking much about music anymore, but about science.

After six months, their Wednesday night lessons ended; but they remained friends, and Snyder became a summer student in the lab of Marion Kies, where Brown worked.

Later, Brown left for the prestigious Pasteur Institute in Paris. Snyder went on to medical school, staying on at NIH as a technician during the summers. By the time he was a senior, doing experiments was second nature to him, and the lab was home to him more than medical school itself.

Marion Kies's laboratory, where Snyder worked, was located on the second floor of Building 10, in the south wing of D corridor, just across the hall from Julius Axelrod's lab.

•

Snyder was born in Washington, D.C., in 1938, the second of five children. He grew up in a middle-class neighborhood in the northwest part of the city. His father was a government cryptanalyst, his mother a contest entrant whose ability to identify mystery melodies and confine her praise of household products to twenty-five words or less won the family, among other things, a trip to Jamaica, a sports car, and thousands of dollars in cash.

Young Sol was the family musician. When he was five, his piano playing won top prize on a radio show called "Uncle Bud's Amateur Hour." He soon tired of the keyboard, though, and turned to the clarinet, then to the mandolin his Russian immigrant grandfather had given him, and finally to the classical guitar, which he's played since. He could have made it as a classical guitarist; at least that's what his teacher, a friend of Andrés Segovia, thought. His mother urged him to study in Italy. But Snyder felt he wasn't good enough. Besides, how did you make a living as a classical guitarist? Segovia did, but who else?

Along with an ear for music Snyder had a head for ideas. When he was a kid, relatives thought he ought to become a rabbi. He studied Talmud in yeshiva, Jewish religious school. He flirted with philosophy, read Nietzsche and Freud, was captivated by the mind and its intricacies. He enrolled at Georgetown University, helping to pay for his education by teaching guitar. He never earned a bachelor's degree, but instead transferred directly to Georgetown's medical school.

His plan—Snyder always had a plan—was to become a psychiatrist.

At medical school, classmate Carl Merril remembers him as "bright and quick, always right at the top of every curve," with intellectual achievement always topmost among his priorities. Being at Georgetown made him feel like a second-class citizen; he was acutely aware that its medical school, while respectable, was a couple of rungs down from a Harvard, a Stanford, or a Johns Hopkins. Once, he and Merril even drove up to Yale for an interview, thinking they might be able to transfer.

Snyder knew he was quicker and sharper than most of his classmates, says Merril, and was willing to help those not so favored, including Merril himself. One time Snyder got him through a much-feared pathology exam, for which they had to be able to identify any of a hundred or more diseased tissues kept in jars around the lab. Snyder had a prodigious memory, had even taken courses in how to memorize long strings of data. Why didn't they, Snyder suggested, link each tissue for which they were responsible with the size, shape, color, and other identifying marks of the jar in which it was kept, and memorize *that.* The large, dark green jar with the chipped base? Cirrhotic liver! The day of the exam, sure enough, Snyder and Merril were finished long before anyone else.

Merril, now a section chief at the National Institute of Mental Health, paints Snyder as something of a wheeler and dealer during his Georgetown days. One time he got himself elected president of the student psychiatry club, which until then had existed in little more than name. "It gave Sol a chance to invite Seymour Kety to the medical school and go to dinner with him," says Merril pointedly, adding, "In due respect to Sol, it was a good thing to do. It brought a good person to the school."

Snyder's competitive streak was strong, according to Merril, and it even surfaced between the two of them. Merril might make a particularly insightful remark and Snyder would jab:

"I'm not going to tell you anything anymore, Carl. You're going to get ahead of me." He said it playfully, yet Merril always wondered whether he meant it seriously. Even today, he's not sure.

On graduating, cum laude, in 1962, Snyder went to San Francisco for an internship at Kaiser Hospital and, a year later, was all set to launch his career as a psychiatrist. But his new wife, Elaine, needed to return to Washington to satisfy her college degree's teaching requirement. Snyder applied for a position in the NIH Clinical Associate program, but none were available.

"Then, lo and behold," says Snyder, fanfare in his voice, "Julius Axelrod had a research associate slot open."

It was freak good luck. The Cuban Missile Crisis, with its threat of war, was not long past. NIH's Research Associate program, which left you doing scientific research instead of lugging a rifle, had scads of sharp young people competing for its handful of positions. Snyder had a good academic record, but from a medical school of only middling rank. "Everybody else," as Snyder tells it in his sometimes hyperbolic style, "was the valedictorian from Harvard." Ordinarily, he wouldn't have had a chance; indeed, Axelrod's slot had long been filled. But then the successful applicant backed out, those on the waiting list had found other jobs, and Axelrod needed someone.

Snyder, meanwhile, had been prowling the corridors of NIH for a job. "Sol had a way of working with and through people," Carl Merril remembers. He was intensely curious, knew everything and everyone. Now he learned of Axelrod's opening, and made his availability known.

Axelrod checked with Donald Brown, who was by this time in Baltimore, with the Carnegie Institution of Washington's department of embryology. " 'He's a bright boy,' " Axelrod remembers Brown telling him.

In 1963, Snyder joined the laboratory of Julius Axelrod. "I owe everything in my professional life to Julie," Snyder would one day tell a hall full of former students and colleagues of Axelrod. "And there are one hell of a lot of people in this room who can say the same."

•

Axelrod started Snyder off with histamine, a hormone released from injured skin and during allergic reactions. It was, Axelrod figured, a natural for Snyder. First, he already had experience with histamine's precursor, histidine, one of that class of chemical constituents of life known as amino acids. Second, he suffered from asthma, whose symptoms antihistamine drugs relieve.

For Axelrod, histamine was an outgrowth of his work with the catecholamines. For almost a decade he'd been using the enzyme he'd discovered, catechol-O-methyltransferase, as a tool to study noradrenaline. COMT breaks down catecholamines by attaching to them a methyl group. Might there exist other such methylating enzymes? Axelrod had wondered. Sure enough, he and Donald Brown had discovered one in 1959. It was called histamine-N-methyltransferase, or HIMT. And now he had in mind a way to use it for measuring histamine.

Snyder's job? Get it to work.

Previous methods had used, as a kind of biological measuring stick, histamine's ability to contract smooth muscles, or else were based on sensitive but laborious fluorescence techniques. Now Axelrod thought he had a better way: It was known that S-adenosylmethionine supplies the methyl group that HIMT grafts to histamine, forming methylhistamine. Well, what if you used hot SAM, S-adenosylmethionine made from radioactive carbon? Measure the radioactivity of the methylhistamine end product and you'd know how much histamine had been present initially. You would, except for one factor: Without knowing how much of the histamine in the original sample had participated in the reaction, you couldn't complete your calculation.

To sidestep the problem, Axelrod hit upon adding a known amount of a second radioactive-labeled compound, tritiated histamine, to the sample. Some of it would combine with SAM—and in the same proportions as the unlabeled histamine did. Then you'd count methylhistamine radioactivity due to the tritium and, separately, that due to the carbon. Roughly, the less contributed by the tritiated histamine, the more (unlabeled) histamine in the original sample.

Today, Snyder enjoys telling how Irv Kopin, who worked upstairs at the time, assured them that their double-labeling technique wouldn't work. Coming from Kopin, that carried weight. "He's smarter than Julie, smarter than me," says Snyder. Kopin assured him he was wasting his time, offering what Snyder concedes was unassailable logic. "He proved definitively that it couldn't work."

It worked. The method worked so well it could detect histamine in amounts down to two billionths of a gram. It opened up research into histamine, says Snyder, and twenty years later, is still in use. "Real dogged biochemistry," Axelrod calls their work. "Snyder claims to be a klutz in the lab, but he knew when to be careful."

•

"In collaboration with Solomon Snyder, we studied the mechanism of the serotonin cycle," Richard Wurtman and Julius Axelrod began a *Scientific American* review of one phase of their research into the tiny, cone-shaped organ embedded deep within the brain known as the pineal. But, in fact, there was no collaboration. Wurtman worked on the pineal. Snyder worked on it. Axelrod worked with both of them. But Snyder didn't work with Wurtman. By all accounts, the two younger men could scarcely tolerate one another.

Axelrod had met Wurtman at a scientific conference at which Wurtman had ascribed the strange behavior of a pigeon, following its injection with adrenaline, to the drug's reaching the central nervous system. Couldn't be, Axelrod said; adrenaline couldn't pass the blood-brain barrier. In fact, the adrenaline, exerting only its normal peripheral effects, had *scared* the pigeon, making Wurtman think the drug acted directly on the brain. Still, the discrepancy had got them talking. Soon, Wurtman—a tall man of astronaut-erect bearing with an M.D. from Harvard—was Axelrod's research associate and the two of them were working on the pineal.

The pineal gland, the only unpaired organ in the brain, for eons remained a mystery to medical science. The ancients saw it as the "third eye." Descartes imagined it as the seat of the rational soul, receiving images through the eyes and regulating the passage of "humors" through long hollow tubes to the muscles. When, in 1965, Axelrod and Wurtman wrote about the gland in *Scientific American*, they noted that only five or six years before, the pineal had been viewed variously as a photoreceptor in frogs, as playing a role in sexual function in rats and in humans, and as containing something that blanched pigment cells in tadpoles. In other words, the pineal mystery was just that, a mystery.

One fact was known: it inhibited the growth of the gonads, or sex glands. But just what in the pineal did that? "Our plan," Wurtman and Axelrod wrote, "was to subject extracts of cattle pineal glands to successive purification steps and test the purified material for its ability to block the induction by light of an accelerated estrus cycle in the rat." It was a job as tedious as the description of it, yet one they had to be ready to tackle if they were set on getting an answer.

They never did it. "We were both pretty lazy," says Axelrod, his way of saying he had a smarter, simpler approach in mind. "We decided to take a flier. . . ."

A few years before, a compound isolated from the pineal had been found to blanche the skin of tadpoles. It had been named *melatonin*, from its effect on the pigment melanin. It was strong stuff; a trillionth of a gram was enough to wash out the color from several square inches of skin. (Maybe, some thought, it would find use in dermatology.) Melatonin's other effects, if any, were unknown.

Now Axelrod hoped to short-circuit grim weeks and months of tedious lab work. Instead of doggedly searching for whatever pineal substance inhibited gonad growth, they would, in vintage Steve Brodie style, guess what it was: "Let's see if it's melatonin," Axelrod said. It was.

Wurtman and Axelrod had been working on the pineal for perhaps a year when they were joined by Sol Snyder. Not long before Snyder's arrival, Wilbur Quay, of the University of California at Berkeley, had shown that serotonin levels in rat pineal went up and down through the day, the highest levels being reached at about noon, the lowest at midnight. Serotonin, recall, is the substance that Brodie, Udenfriend, Herb Weissbach, and Park Shore had, during the fifties, helped show is a brain neurotransmitter. More recently, Weissbach and Axelrod had found another role for it, as a metabolic precursor of melatonin. Now Axelrod hoped to look into Quay's through-the-day serotonin rhythms.

But a major hurdle obstructed any approach to the problem: There existed no method sensitive enough to measure serotonin in minute quantities. True, serotonin was highly concentrated in the pineal. But even so there wasn't much of it to measure. Even in a two-hundred-pound man, a pineal weighs but a tenth of a gram; a rat pineal is barely visible to the naked eye. A 1956 paper out of Sidney Udenfriend's group had contributed a spectrophotofluorometer-based method sensitive enough to measure serotonin levels in the brain. But the rat pineal is so small—thirty thousand to the ounce—that dozens would be needed for even a single measurement. Tracking serotonin levels through the day, and under a variety of conditions, then, was close to impossible—not out of theoretical considerations, but simply logistical ones.

Snyder isn't sure just how they became aware of the paper by J. W. Vanable that gave them the clue; he thinks Axelrod may have seen it before it appeared in the literature. In any event, Vanable had found that if you heat a mixture of serotonin and ninhydrin in water, the resulting product is highly fluorescent. Ninhydrin is the ubiquitous laboratory chemical that betrays microgram quantities of amino acids by forming an intense blue hue. "I started messing around with this," remembers Snyder, and soon he and Axelrod had refined a method ten times more sensitive than any then existing: A serotonin measurement once requiring twenty pineal glands now needed but two.

New method in hand, Snyder and Axelrod confirmed that pineal serotonin did indeed go through daily cycles opposite to those of melatonin. Did these daily cycles result from the animal's response to day-night light cycles? To answer that question, Snyder and Axelrod blinded a rat, expecting to find that serotonin's ups and downs disappeared. They didn't. Levels at noon were still ten times higher than at midnight. "I could hardly believe it," remembers Snyder. They'd found, it seemed, a biological clock, synchronized by environmental lighting, but otherwise independent of it. Only a diet of continuous light stopped it.

A whole series of experiments followed. They found that rats plunged into a reversed day-night cycle could acclimate, their serotonin rhythm locking into the new cycle within six days. They learned that the pineal was supplied by nerves that release noradrenaline. They came to see the organ as a neurochemical transducer, translating light energy reaching the eyes into hormonal secretion. Bit by bit the veil obstructing an understanding of the pineal gland lifted.

And all the while, Snyder and Wurtman clashed. "When Wurtman did X, Sol had to do Y," says one knowledgeable of their rivalry. "They didn't talk, they yelled at each other," Axelrod remembers. He felt it was pointless to intervene. "I didn't want to be bothered by it," he admits. "I could try, but I knew I couldn't do anything about it. I'm not a psychiatrist."

Irv Kopin, in and out of the lab during much of this period, remembers Wurtman as aggressive to the point of arrogance. Snyder, by comparison, was easygoing, "a loose person, easy to interact with, enthusiastic, always eager to try something new. He reminded me of Julie." But like Axelrod, Snyder, "didn't dot every *i* and cross every *t*," as then-lab chief Seymour Kety puts it. Often, he had trouble spelling out his reasons for doing things. That inflamed the coolly analytical

Wurtman. "You have no idea what you're doing!" he once blurted out to Snyder in frustration.

Wurtman would get similarly frustrated with Axelrod, but the older man tolerated it better than Snyder. Besides, as Kety says, "Julie was very enamored of Wurtman." Wurtman had introduced him to the pineal gland and, by the time Snyder appeared, "was already established in Julie's affections."

Kety feels that Snyder was probably the brightest postdoc Axelrod ever had. Axelrod remembers young Snyder as curious and eager, "with a certain flash of brilliance" and an almost frighteningly quick mind. Yet Kopin, for his part, feels Wurtman may actually have been the smarter of the two, with Snyder the more creative. "Dick reasoned out everything very carefully, while Sol was more intuitive. They were like two children with different personalities, with Julie feeling parental toward both of them." And each of them, in turn, feeling properly filial toward him.

Snyder describes Axelrod as "a wonderful mentor, probably one of the most creative scientists in the world," and his time at NIH as a "golden period" for NIH and for him personally. Today, recalling those heady days in Axelrod's lab, joining it at a time when it fairly shimmered with energy, Snyder can hardly say enough about them—"exciting!" and "wonderful!" and superlatives of every stripe pepper his recollections. "Nobody was there to do anything but research. You were discovering things, and Julie would get excited, more excited than you were. . . .

"Perhaps the greatest lesson Julie taught was that science is fun and exciting," Snyder once said in a formal tribute to his mentor. Snyder pictures him leaning over the scintillation counter, urging it on to the hoped-for figure, like a teenager at a pinball machine. Axelrod could see in apparently trivial data whole worlds of scientific possibilities, yet would advance his ideas "in such deceptively simple ways that, at first glance, they seemed incredibly naive."

Snyder knows that some found Axelrod something of a nag, assigning you, for example, an experiment likely to take three days, then showing up that afternoon after lunch to learn whether you'd turned up anything. But Snyder loved it. "It gave me a chance to say, 'Yes, Julie, let me show you.' "

Axelrod made young Snyder feel the work they did together was the most important in the world. He taught him that "science is just as creative as any of the arts," would talk of theories that were "beautiful, symmetrical, the kind of thing you got excited about, lost sleep over."

Today, as Snyder recounts those early days with Axelrod in 2D45, on one wall of his office hangs a framed photograph bearing the inscription "Dear Sol," and signed "Julie Axelrod": "Thank you for your help in making this day possible for me," it reads. The photograph is of Nobel House in Sweden, and it's dated December 10, 1970, the day Axelrod received the Nobel Prize.

Axelrod, says Snyder's friend Carl Merril, was like another father to him. After two years in Axelrod's lab and two dozen published papers, Snyder came away infected with what he'd call the "virus of research." He had been born as a scientist. He would devote his career, he decided, not alone to talking with troubled patients, but to unraveling the chemistry within their troubled minds.

•

Once or twice during his two years with Axelrod, Snyder met Steve Brodie. He'd later learn to appreciate Brodie as the father of drug metabolism, as a pioneer in neuropharmacology, as mentor to many of the field's most towering figures. Back then, though, all he knew of Brodie was what he heard from Julie: " 'That son of a bitch,' he'd say." As for his own contacts with Brodie, they were cordial but brief: "I was a little squirt. He was a big shot."

During the late 1950s and all through the 1960s, Brodie had become an international scientific figure of vast reputation. The awards, the visiting professorships, the medals and honorary degrees had started coming his way about 1960, and all through the decade they continued. He was made a Fellow of the American Association for the Advancement of Science in 1960, a Fellow of the New York Academy of Science the same year. He received the Shionogi Commemoration Lecture Award in 1962, an honorary doctor of science degree from the University of Paris, and the Torald Sollmann Award in Pharmacology in 1963. He was made an honorary member of the Czechoslovak Academy of Sciences in 1965. Later, he'd be named to the National Academy of Sciences, be awarded an honorary doctor of science from the University of Barcelona, receive the Lasker Award, accept the National Medal of Science from President Johnson—on and on it went.

The expansion of his scientific reputation and the resulting influx of foreign scientists to the lab had begun in the middle and late 1950s. Back in 1956, for instance, Marcel Bickel, a graduate student at the University of Basel, in Switzerland, had been handed a doctoral thesis problem on barbiturate metabolism. Knowing little about drug metabolism, he set off for the library to read up on it. There he first encountered the name of Brodie.

Five years later, he was a member of Brodie's lab himself. "I arrived one day in 1961," he reminded Brodie later, "overwhelmed by the spring blossoms of Bethesda, by the colossal complex of NIH, and by an initial discussion with you that lasted for hours." (Their talk took place in the middle of the day, he added, "a rather remarkable exception, as things turned out.")

Typically, Brodie got into the lab about noon, reporting to his paper-cluttered office in 7N117. Checking in with his secretary, he'd note his appointments, make phone calls, then spend most of the afternoon talking science with his lieutenants, trading ideas, suggesting experiments, checking up on what everyone was doing. Lunch was apt to be a brown-bag affair at his desk by the window. At six or so, Mrs. Brodie would pick him up and drive him home for dinner. Then, around eight or nine, colleagues would begin filtering into his apartment, which was really an extension of the lab, to go over the latest data, discuss ideas, and write papers.

Brodie credits James Shannon with instilling in him the importance of good scientific prose, and he took equal pains to inculcate sound writing principles in his students: Make free use of the thesaurus, he'd advise them. Avoid the passive voice. "Let the verb do the work." He loved words, says Park Shore, remembering how Brodie was always getting up to consult the dictionary. "His papers just flow beautifully."

Brodie was a salesman of his ideas. He demanded that his papers not simply supply the facts but be readable. "That's one reason he was as famous as he was," says James R. Gillette, who ultimately succeeded Brodie as lab chief. "He could make understandable the complex. And he went in for little phrases that would perk the imagination."

A paper bearing the imprint of the lab went through endless drafts before Brodie, going over it line by line and word for word, peering back at you over the tops of the reading glasses he usually wore, was satisfied. Lewis Schanker, who joined Brodie's lab in the 1950s, remembers how the first manuscript he got back from Brodie was so dense with hieroglyphics he could scarcely find what he'd written underneath.

Barbara Orlans, another veteran of those late-night sessions at Brodie's apartment, tells how the real work of the evening got under way only once the first draft of a paper had been chucked into the waste basket. Then, "closeted with the boss, we would puzzle over a word for ten minutes, check the dictionary for synonyms, rephrase the whole sentence to avoid that particular word, and then throw the whole paragraph out. The hours went by, dates and engagements were cancelled, midnight came and went. . . ."

The evening shift at Brodie's place could expect to find sharpened pencils, food, and the hospitality of Anne Brodie. "She was our good angel," recalls Alfred Pletscher. "Thanks to Anne, our work could develop in a relaxed and peaceful atmosphere." The same could be said of her influence on Brodie's work all through his NIH years.

He'd met her in the late 1940s, when both lived in the

Beaux Arts Apartments in New York. She was a striking blond just past forty, a secretary from up on the twelfth floor who'd taken to accompanying an elderly woman from the building on walks. Returning from one such stroll, decked out in a big hat and white gloves despite the August heat, she met Brodie in the lobby. Soon they were seeing each other regularly. On August 31, 1950, they were married.

(Brodie had been married before, to a woman named Frieda Harris, but he shrugs off the relationship as lasting but briefly. It formally ended in divorce in 1939, and even by the early 1940s, according to one who knew him then, "Brodie was acting as though it had never happened.")

Anne Smith had grown up as one of six sisters on Waverly Place in New York's Greenwich Village, at a time when horsedrawn streetcars still plied Manhattan streets. Her father, a Russian literature enthusiast at the time she was born in 1905, had named her Anastasia; the kids laughed at her and soon she was just Anne. When she was older, her Buster Brown haircut got her a boy's part in a show in which another sister also played, and for a while she did one-night stands on the vaudeville circuit, at the end of each matinee hopping a train to the next city. She retired, she's fond of saying, when she was eight.

Anne was not the "nice Jewish girl" Brodie's mother had wanted her son to marry. Still, after his father died and Brodie brought his mother to New York, she and Anne grew close. Anne remembers Mrs. Brodie appealing to her: "If I die, will you take care of him for me?"

She did soon die, and Anne did take care of him. Oh, she'd grumble about his fondness for meat and potatoes when she was inclined to gourmet fare; and about his nocturnal work habits—though never to any effect. But he remained always, to her, "The Doctor." She would buy his airplane tickets, pack his bags, stay up all night typing draft after draft of his papers, keep up the scrapbooks that recorded his honors, awards, and appearances in print. She doted on him, "waited on him hand and foot," as Gene Berger recalls of their early relationship. Some read in their marriage elements of a mother and child's mutual clinging, or even, less charitably, a master and servant relationship. But Anne Brodie herself, an intelligent, well-read woman with firm opinions on all manner of social and political issues, explains it less tortuously: "I'm devoted to him."

She calls the years during which Brodie reigned over the Laboratory of Chemical Pharmacology "the golden era." At its height, Brodie was like a scientific potentate, traveling all over the world for conferences, his waking hours given over to unraveling nature's mysteries, not to fretting over mundane travel details. One time, on the way to Washington's National Airport, his friend Costa remembers, the two of them got to talking about an experiment, missed a turn, and wound up driving aimlessly through Arlington Cemetery.

Another time, in Geneva, Brodie was supposed to be boarding a flight to New York when, distracted by his musings, he marched past armed soldiers to the wrong plane. No one stopped him, he took his seat, the plane took off—and the pilot announced their arrival time in Moscow. (Brodie summoned a stewardess and the Aeroflot jetliner returned to Geneva.)

While in Washington, Brodie's favorite haunt was Blackie's House of Beef, a downtown Washington landmark frequented by celebrities whose glossy pictures lined its walls, where he could get a thick slab of roast beef, salad, and baked potato with sour cream and chives for four dollars. Brodie would go there armed with books, papers, and pencils, and camp out there for hours. All the waiters knew him.

One time, in the 1950s, he struck up a conversation with a busboy. The busboy held a bachelor's degree in chemistry. Brodie offered him a job. "Mrs. Brodie became my 'adopted mother,' [and] you became my father and guide for the rest of my life," Peter Neff wrote him in thanks years later. While they ate TV dinners or watched "Gunsmoke" at Brodie's apartment, Brodie would tutor him, pushing him, trying to make a researcher out of him. It didn't take. But Brodie's example and faith supplied powerful motivation, and Neff went on to become a professor of dentistry at Georgetown University.

In the lab or out, Brodie enjoyed himself, delighting in the free play of his opinions and ideas, exulting in his own spontaneity. Brodie once taped a baseball game for Mimo Costa's son. Whenever the excited announcer screamed that one or another big-league star had hit a home run or stolen a base, Brodie would voice over the major leaguer's name with that of Costa's son, to the boy's endless delight.

For a time, Brodie, appearing under a pseudonym, took to writing mock-serious articles on pseudoscientific topics for a medical newsletter. In one column in the series, which was dubbed "Artefacts and Fancies," he went the standard double-blind methodology used in drug testing one better with a "triple-blind" test: "The subject doesn't know what he is getting, the nurse doesn't know what she is giving, and the investigator doesn't know what he is doing."

In another of the series, he reported on "research" suggesting that bananas had a nervous system: "High doses of LSD evoked psychotic behavior manifested by indifference to usual physical laws. In a heavy storm the trees bent *against* rather than *with* the wind, indicating a loss of contact with reality." And similar such charming silliness.

Brodie took special delight in pricking bureaucratic balloons. A jargon-laden government memo would cross his desk and he'd fire off a note of protest to the heart institute's longtime administrative secretary, Evelyn Attix, who ultimately clipped together a sample of his correspondence under the title, "A Series of Documents Supporting BBB's Championship Status Relative to Torturing Administrators."

In July 1962, for example, Brodie got a particularly indecipherable memo from an NIH deputy director and sent it back to Attix literally gray with penciled-in notes: "I strongly urge that this interesting memo be translated. I know it must be important since I do not understand a word." On another occasion, Attix sent Brodie an example of how an NIHer had graciously returned a twenty-five-dollar honorarium he'd mistakenly received for a lecture. Bah, Brodie shot back: "Dishonesty varies inversely as the square root of the temptation," and he proceeded to perform a mock calculation proving that there was little to laud in the action.

Still another time, Attix asked Brodie whether he wished to continue receiving certain government publications. No, he replied, "anything you can do to stop the torrent of inconsequential gibberish will be appreciated," he wrote back. "*We must save our forests.*"

It was the prime of Steve Brodie.

•

Meanwhile, down the hill in Building 1, James Shannon had successfully transformed the National Institutes of Health into an instrument of his scientific vision.

Twenty years before, at Goldwater, Shannon had been seen as "waking to a new and not unsympathetic capacity in himself for wheeling and dealing." Now, as head of NIH, that capacity had come to fullest flower. "A little [like] Santa Claus on one hand and a little Machiavellian on the other," is the way somebody once described him. Santa Claus in how he backed up his people, getting them whatever they needed; Machiavelli in how he got from them whatever *he* needed.

Each year Shannon would go before Congress and appeal for money. He was no orator; he could put people to sleep during those budget hearings, his tendency to mumble being no help. (Indeed, it's been said, presumably in jest, that the reason Shannon was so successful at extracting money from Congress was that no one ever knew just what he'd said.) During the questioning that followed, Shannon was better, coming back with quick, authoritative, no-nonsense replies that sometimes left his interrogators spellbound.

But it was before he ever walked into the committee room that Shannon was at his best. "He had the most amazing kind of personality I've ever encountered," someone once said of him. "His knowledge of science was awesome. Yet it was as if he had only to take off his jacket to become a roaring Irishman." In informal settings, around a table with just a few others, he showed real charm and wit, and was most in his element. And sitting around a table was just what he did each year before those appropriations hearings, where he'd work out the NIH budget with his friends Senator Hill and Congressman Fogarty.

On the NIH campus today, Building 16 is called the John E. Fogarty International Center for Advanced Study in the Health Sciences. Building 38A, a 166,000-square-foot high-rise office tower, is the Lister Hill National Center for Biomedical Communications. Each represents NIH's way of immortalizing congressional leaders—Fogarty was a Rhode Island congressman, Hill a senator from Alabama—who did much, during Shannon's term as director, to funnel money NIH's way. "These people were all strong and good friends of the NIH," Dewitt Stetten would note at the time Building 1 was renamed the Shannon Building in 1983. "But the real conspirator, the leader of this group, was Jim Shannon."

His success was phenomenal. By the hundreds of millions, then by the billions, research dollars poured onto the NIH books. In the five years before Shannon took over, the NIH annual budget had crept from $52 million to $81 million. In the next five years, it quintupled to $430 million, and by the time Shannon left in 1967, it stood at $1.4 billion.

Shannon orchestrated his appeals on behalf of NIH with what longtime aide Bill Carrigan calls the "three-legged stool" approach. The three legs—research, training, and physical plant—were the necessary supports on which the NIH mission rested. "Naturally," as Carrigan says, "one leg was always short." That is, Shannon would profess satisfaction with two of the legs and paint the third as dangerously weak. Then, in later years, with the robust growth of the underdeveloped leg, he could legitimately claim that one of the others had become, by comparison, short.

Formally, the budget presented to Congress for review is the president's budget. But Shannon usually got Congress to add substantially to it. Once, the story goes, the budget process was over for the year and Shannon was at an official function with President Eisenhower. Loosened up by a couple of martinis, Shannon reportedly slapped Ike on the back and said, "Well, boss, looks like we beat you on the budget again this year."

More typically, though, Shannon was all seriousness. He was usually up at six in the morning, and worked for a couple of hours at home before coming to the office. Early-morning meetings with him, Dewitt Stetten remembers, left scant room for lightness. And he could be short with you. Ask what he deemed a foolish question and he'd get irritated. "He was," in Park Shore's recollection of him, "a remote and rather awesome character, with a crystal clear, driving mind."

His insistence on personal excellence helped NIH firmly shed the prevailing image of government laxity and incompetence. He had no patience for mediocrity. One otherwise ardent admirer complains that Shannon saw only black and white; you were either good, or you were not. "If you performed," says Tom Kennedy, "he called on you to perform again and again. Otherwise, he just forgot your name."

In *The Youngest Science: Notes of a Medicine-Watcher*, Lewis Thomas wrote of what he saw as the great change in medical education and research that came with the expansion of NIH after World War II. During the mid-1950s, Thomas was for a time a member of the National Advisory Health Council, which was supposed to help set NIH policy. "We had the time of our lives," he wrote. "Everything seemed possible." Congress was high on research. Medical schools were of a mind to expand their research facilities. Money flowed freely. "And Dr. James Shannon, the director of NIH, knew exactly where he wanted NIH to go and how to lead it to its destiny. . . .

"In retrospect, it can be seen that expansion of NIH and the recruitment of medical faculties for implementing the national mission of NIH represented one of the most intelligent and imaginative acts of any government in history, and

NIH itself became, principally as the result of Shannon's sheer force of will and capacity to plan ahead, the greatest research institution on earth."

•

It was 1965 and Sol Snyder, then twenty-seven, was leaving NIH. He'd planned to become a psychiatrist and still wanted to get a psychiatric residency under his belt. Yet after two years with Julius Axelrod, he knew he wanted to do research, too. Could he do both?

Stanford University, he learned, had just the kind of program he sought: You spent time in the lab, but also saw patients. And you got more than the lousy 250 bucks a month or so that psychiatric residents usually got. "That's where I want to be," thought Snyder.

But Stanford had no lab space for him. And by the time he found out for sure, most other desirable residencies were filled, too.

He went to see Axelrod. Sure, he could stay another year if he wanted, Axelrod told him; so at least the pressure was off.

He went to see Seymour Kety, who thought that maybe Johns Hopkins in Baltimore could strike a Stanford-style deal with him. But when Kety actually looked into it for him, he reported back that no, Hopkins couldn't do that. But would he like to come anyway, Hopkins wondered, as a regular psychiatric resident? Snyder said he'd let them know.

He traveled to Cleveland for an interview at Western Reserve University, met some of the people in its pharmacology department, gave a lecture, made a good impression all around, and came away with the offer he'd wanted—a psychiatric residency plus an assistant professorship in pharmacology.

He returned to Washington and called Hopkins to say that, no, it looked like he wouldn't be coming, that he'd gotten an offer . . .

Oh, they'd already heard about the Western Reserve offer, the man at Johns Hopkins said, and had decided they could make a similar deal, after all. In fact, says Snyder, smiling, plainly relishing the story, they could do better yet, with more money, and some other nice perks.

That was more like it, thought Snyder. Next stop: Baltimore.

9. Johns Hopkins

SHE MET HIM at the foot of the towering white marble statue of Jesus Christ, under the great dome that is the century-old symbol of Johns Hopkins Hospital. It was raining, and Solomon Snyder, umbrella in hand, had just whisked down from his lab in the Wood Basic Science Building to welcome her to Hopkins. Even as they made their way along the hospital's teeming corridors, exchanging pleasantries, she felt "inundated by the sounds and the busyness of Hopkins. There was this incredible feeling of intense energy, this frenetic *ztzztzz*."

Her name was Diane Russell. She was thirty-two and had just earned a Ph.D. from Washington State University. She was here in Baltimore to begin a postdoctoral fellowship under Snyder, who, though three years her junior, was already an assistant professor of pharmacology.

As careers in science go, Russell's had gotten off to a late start. Born and raised in small-town Idaho, she was twenty-four and a mother of three before she went to college. At Boise Junior College, immediately drawn to the sciences, she blossomed. After receiving an associate of arts degree, she went for her bachelor's at the College of Idaho, a Presbyterian institution of small size but fine reputation. She graduated, summa cum laude. Then it was on to Washington State for graduate school, where Donald S. Farner, a disciplined, rigid, and Germanically thorough ornithologist who was working on circadian rhythms in birds, oversaw her methodical march through the university's Ph.D. requirements.

It was September 1967 when she got to Johns Hopkins, its hospital and medical school stuck squarely amidst block after block of two- and three-story, marble-stooped row houses. This was where Russell wanted to be. She knew that the academic elite to which she aspired were largely drawn from graduate and postdoctoral programs at top universities back east—like Johns Hopkins. She'd applied for a postdoc with Guy Williams-Ashman, a top Hopkins endocrinologist. He, having no lab space available, passed her application on to Snyder.

Snyder was outside her field, and just starting up. It was like going to work for an unknown, remembers Russell. "But I banked on the fact that if he were at Hopkins he would be good enough to learn something from."

Snyder, who'd arrived in Baltimore two summers before, was both a psychiatric resident, seeing patients in his office, and a pharmacology researcher. And already he was beginning to build up his fledgling lab. Edith Hendley, then at Hopkins's Wilmer Eye Clinic, had followed with rapt interest the revelations about sympathetic nerve function emerging from Julius Axelrod's lab at NIH. His was "the most exciting neurochemistry lab anywhere," and that Snyder was its product was good enough for her; she became his first assistant. Three young medical students followed. Now, in Diane Russell, he had his first postdoc, willing hands for his ornithine decarboxylase idea.

One of Snyder's scientific bailiwicks, from NIH days, was histamine metabolism, and a recent paper out of Sweden had caught his eye: The enzyme that changes histidine into histamine, histidine decarboxylase, apparently played a key role in the feverish cell growth that accompanies the healing of wounds. Could it be, Russell remembers Snyder suggesting, that, in other instances of rapid cell growth, amines other than histamine might be involved?

Several lines of evidence pointed to the polyamines, a class of compounds marked by amino groups at both ends of their chainlike structures and names, owing to peculiarities of their discovery, of distinctly unsavory cast. Chemically, their starting point is the amino acid ornithine—which, helped by the enzyme ornithine decarboxylase, becomes putrescine, one of the polyamines, and precursor to the others, spermidine and spermine. Putrescine was originally isolated from cholera bacteria. Spermine was first found in human semen three centuries ago by Anton van Leeuwenhoek, inventor of the microscope.

Conventional wisdom had consigned to the polyamines a role only in bacterial processes. But just before Russell's arrival in Baltimore, an article in *Science* had reported the presence of putrescine in regenerating rat liver (which, when part is sliced away, grows back to its original mass, like a salamander's leg). Liver regeneration means rapid cell growth; so does wound healing—where histidine decarboxylase had been found the "rate-limiting" enzyme. Perhaps, thought Snyder, noting the *Science* results and reasoning by analogy, ornithine decarboxylase (the putrescine-making enzyme) played a similar role in liver regeneration.

It was a stab in the dark, a "flier." But on the rainy September day on which they first met, this was the possibility Snyder and Russell discussed. Within two weeks, Russell was cutting up rat livers.

The experimental plan: Anesthetize the rat. Slice out part of its liver, stimulating the regeneration process they wished to study. After a few hours, kill the animal. Remove the liver remnant and grind it up. Centrifuge out the solid parts of it, leaving behind, among other things, the enzyme that makes putrescine, ornithine decarboxylase. To the mixture add ornithine, its carboxyl group labeled with radioactive carbon.

The idea, then, was that when ornithine's carboxyl group

got lopped off, by ornithine decarboxylase, it would form hot carbon dioxide, which they could then collect and check for radioactivity on a scintillation counter. If ornithine decarboxylase did indeed play a role in rapid cell proliferation, radioactive-labeled carbon dioxide ought to collect in abundance.

Immediately Russell ran into a problem: When she anesthetized her laboratory rats with a regular dose of pentobarbital and removed two thirds of their livers, the rats refused to wake up. Day after day, they remained asleep. The problem perplexed her until she recalled what Julius Axelrod and Steve Brodie, among others, knew intimately—that it was the liver, with its microsomal enzymes, that metabolizes drugs. With most of the liver gone, the pentobarbital didn't get broken down, and so continued to act. Ultimately, Russell substituted ether, a gas that scarcely reaches the liver.

After that, results came with satisfying speed. As a basis of comparison, one group of rats had undergone a "sham operation" that left the liver untouched; among these rats, therefore, no unduly rapid liver cell growth was expected. When, four hours after the operation, they were killed and the rest of the experimental protocol carried out, the carbon dioxide reading was, sure enough, a low 3.2. For the rats whose livers had been partially removed and thus were busy regenerating, the reading was 38.9—eleven times higher. Sixteen hours after the operation, the difference was even more pronounced.

Nothing subtle about it. No interpretation needed. Just a single black-and-white result that proved their hypothesis with stunning, graphic clarity. For Sol Snyder, all that radioactive carbon dioxide bubbling up was the kind of exciting lab result he'd experienced many times before. For Russell, it meant much more: "It hooked me on science forever. It made me into a science junkie."

Russell and Snyder confirmed their initial finding in chicken embryos and tumors, both of which also exhibit rapid cell growth. Their paper, "Amine Synthesis in Rapidly Growing Tissues: Ornithine Decarboxylase Activity in Regenerating

Rat Liver, Chick Embryo, and Various Tumors," appeared in the *Proceedings of the National Academy of Sciences* in 1968. In the next fifteen years, it was cited in more than 630 other papers, helping to open up a whole new field, in which Russell is today an acknowledged leader. Polyamines have turned out to be key markers of pathological conditions, including cancer. Whole conferences are devoted to them. Today, in her lab at the University of Arizona in Tucson, a sign outside her office proclaims POLYAMINE CAPITAL OF THE WORLD.

That first experiment, Russell says, "was a fortuitous fling into the what-if," for which she assigns Snyder full credit. Linking histidine decarboxylase with the polyamines? "It was brilliant," she says. "Sol will pull things from here and there," she says, pulling at imaginary objects around her desk, "to ask a further question. That's the excitement he generates."

Russell's experiments with Snyder represented a sharp break from the style of her first mentor, Farner. Her Washington State Ph.D. advisor had been meticulous and careful, taking one step at a time. She likens his approach to a nerve impulse "that crawls along the neuron. Sol's, on the other hand, was like the sudden jump of synaptic transmission."

•

He has a long narrow face with thin, translucent skin, and still eyes that seem to listen rather than to see. He stands, hunched over, back arched, his large head cocked to the side, as if straining to hear. Or else he sits, head cradled in hands, peering at you through an elaborate superstructure of intertwined fingers. His manner is courtly, his speech measured, smooth, and slow.

That, at least, is Sol Snyder at rest. For in the next moment, the coiled spring abruptly unsprung, the stillness shattered, it is as if Snyder had changed into someone else: borne away by an idea, perhaps setting out the sequence of thought leading to an experiment, or taking two or three parts in a recollected dialogue, he's all over the place, caught up in one or another intellectual drama of his own making, wildly animated, his voice occasionally cracking into a higher register, manic energy boiling over.

His friend Carl Merril recalls that even back in medical school his mannerisms marked him: He was forever scratching the back of his head during a lecture, or looking at you through his fingers while intently listening. Edith Hendley pictures him rubbing his nose, twitching, cursing, pacing up and down.

"Try talking to him when he's standing up," warns another former student. "If you're sitting in the middle of a room, he'll walk around you. The only thing to do is put your back against one wall, so he can only do a half circle. That way you won't get dizzy following him."

Frenetic, with tremendous energy, and so, sometimes a strain to work with—that's how Diane Russell recalls Snyder. "Sometimes he was so frenetic I was afraid he was going to fall apart. You'd have to focus him on the work at hand. He was always rushing to take an experimental result one step further, always asking, 'What's next?' Sometimes it was just too much. Sometimes you just wanted to go to Palm Beach and suck your thumb for a few weeks."

Even in the early days, Snyder was regarded as an up-and-comer, and people invariably came away impressed by the quickness of his mind and the fertility of his imagination. But he was still young, and his reputation extended barely outside East Baltimore. Often, his papers came back rejected. " 'Oh, they just don't understand what we're going,' " Diane Russell recalls him saying. He'd analyze why, revise the paper accordingly, and ship it back out. Failure didn't sap his enthusiasm—not for long, anyway. If you're doing anything important, she learned, you have to live with rejection.

The lab was then a modestly-scaled operation, just two rooms on the third floor of the Wood Basic Science Building with a sort of bridgelike area between them. For a while, Snyder maintained his unusual dual role as psychiatric resident and researcher, spending a couple of days in the lab and then, while his technician carried out his instructions, returning to patients. He enjoyed seeing patients, but enjoyed research more. Gradually, his patient load dropped until it was down to just a few afternoons a week.

Even at the beginning Snyder never did the actual bench work. Later, after he became famous, the Hopkins public relations office issued a publicity shot that showed him in a white lab coat, sitting by a big electron microscope, pencil in hand, notebook at his side, apparently caught recording experimental data. The photo fits the public perception of the scientist, but in Snyder's case it couldn't be more mistaken; after coming to Hopkins, Snyder scarcely went near a test tube. "I'm clumsy," he says, painting chemical explosions and shattered glassware as the inevitable result of his doing otherwise.

A charming pose, says Julius Axelrod, who remembers him taking extraordinary pains with experiments when necessary. Yet perhaps not entirely so. One former student, Gavril Pasternak, recalls how once he was getting ready to inject some rats, a job in which one person normally holds the rat, the other the needle. It was seven in the evening and everyone else had left for the day, so Pasternak peeked into Snyder's office and asked for a hand.

"You want to inject a rat?" said Snyder, grabbing the animal. "*I'll* show you how to inject a rat. I'll show you how we used to do it in Julie Axelrod's lab!" As he relates the story, Pasternak's eyes assume the wild-eyed look, his voice the squeaky nasal twang, that is Snyder's at his most animated.

"You take . . . the rat," Pasternak has Snyder saying, all manic energy. "And you take . . . the needle. . . ."

But Snyder was holding it wrong, the rat squirmed free, reached around, and bit him. "Goddamned rat!" shrieked Snyder, throwing it against the wall. "That was the last time I ever asked Sol to help me with an experiment," says Pasternak.

Snyder tells of a distinguished Hopkins researcher, one much older than he, who still works at the bench on his own experiments. "He loves it," Snyder says. "But it's just not my style. It's a bad use of my time." Better use of it, he feels, is to conceive many experiments, for many students, thus magnifying his impact.

Which, of course, demands a reservoir of ideas deep enough to keep them busy. In this respect, say former students, Snyder had no peer, ideas rolling out of his head like silver dollars from a Las Vegas slot machine. "His mind knew no bounds," says Edith Hendley. He was at home in fields far from his own, and could relate them to the work of his own lab. "He knew molecular biology, played with molecular models the way a physical chemist does," according to Hendley. "He's a true genius." The average lab chief might be able to follow three or four research projects at a time. He could handle a dozen students, and remember what each had found six months before.

His mind roamed over every area of the scientific literature, unshackled by disciplinary boundaries, reaching here, backpaddling into a seemingly unrelated area there. "There's a tendency," says Snyder, "to say, 'I'm a neuroscientist, so I can't study cancer, or the liver.' But I'm not like that. It's *all* exciting. It's all the same to me."

Robert Goodman, who joined Snyder's lab while still a Hopkins undergraduate and retained links to it while working toward M.D. and Ph.D. degrees, tells how in an immunology or genetics article, say, Snyder might find the spark of an idea. "Maybe we could try something like this," he'd scribble into the margin, circling the relevant section and sticking it in Goodman's mailbox. It happened several times a month.

There was constant interaction, Snyder always wanting to know how your work was progressing. Not for him the Isolated Lab Chief style; rather, he worked with and through his students. Says one of them, "The idea of the lone scientist muttering Latin in the corner—that's just not Sol."

And unlike some scientists, he was thoroughly at home in social situations. One former student describes him as having broader interests than any scientist he's ever met. "You can have a conversation with him about anything." Says another,

"I would sooner spend an evening with Sol than just about anybody." Often he had welcome dinners for newcomers to the lab at his Rogers Avenue home in northwest Baltimore: dinner, drinks, lively talk. A recent graduate of the lab, Robert Gould, now with Merck, Sharp, and Dohme, the drug company, recalls him as affable, gracious, adept at putting his guests at ease.

Around the lab, he would hoist himself onto a lab bench to chat with a student about his or her project. He got along with everybody, remembers Edith Hendley. Technicians and secretaries loved him. He had a kind word for everyone. Adele Snowman, a technician who joined Snyder in the early 1970s and whose awesome efficiency in the lab is legendary, calls him probably the best boss she's ever had. Snowman left the lab in 1978 when her husband got a job in another state, but returned later. "He lets me be who I want to be," she says. "He gave me a chance to develop."

Former students describe Snyder as a skillful handler of people, adept at knowing when to issue and withhold praise, and possessed of rare savvy in getting the most from them. Says one, "He had a way of doing these carrot-and-stick combinations that were tailor-made for each student." Rarely did he get angry, or even express disappointment. "You'd do it to yourself," says Robert Goodman. "Everybody is there competing for who gets patted on the head by Sol." One woman student remembers working late one evening, capping vials for the scintillation counter, when Snyder walked in. "Oh, my little baby! What are you doing here so late?" he exclaimed. Even years later, it was plain, the memory was a treasured one.

Good lab data left Snyder thrilled, his enthusiasm brimming over. " 'Gee, you're wonderful. You're the most brilliant person in the lab,' " Robert Gould remembers Snyder telling him at such times. "Of course, tomorrow somebody else would be the most brilliant person in the lab." Still, you'd come away from a half hour with Sol feeling as if you'd solved the riddle of the universe.

When Gould first came to the lab as a postdoc, he was struck by the friendliness of his new lab chief, his sensitivity to personal needs. In time, though, he came to feel that some of it was cultivated, that Snyder knew the impression he wanted to make, and worked at making it. "Snyder is terrifically calculating," says Gould. "I think he's the most calculating person I know."

It is a sentiment echoed, in one form or another, by many who have worked with Snyder. All agree he is warm, supportive, generous. But some question his motivation, seeing his endearing personal traits as manipulative, his skill at getting the most from people just that—a skill, one he works at refining. "I always felt the warmth was to reach ends," says Diane Russell, from the early days of Snyder's lab. "He was very effective in dealing with people and the warmth and gentleness were a way of doing that."

But however calculating Snyder may be, Robert Gould, for one, grants that "most of the time it's for the mutual benefit of both of you. 'Gee, that's nice work,' he'll say. 'But we can't publish it yet. Why don't we go for more?' Or, he'll say, 'You know there's this big meeting coming up, in Puerto Rico. Too bad your work is not ready to present. . . .' " At first, one was apt to be angry. But in the long run, his tactics pushed you to do your best.

Gould describes Snyder as endlessly accommodating, as someone you can call up any time and know that if he can help, he will. Result? "A network of people in his debt." Gould hears the Machiavellian note in his comment, insists he means nothing by it. But facts are facts: "There are a lot of people who feel that because Sol has helped them, they want to help him."

Indeed, many see Snyder as the consummate political animal, trading favors and information, his pulse on the centers of power. Snyder knows *everybody* in the councils of big science, it is said, and cultivates friends in places low and high. An acquaintance of fifteen years, one given to harsh, monochromatic assessments, says, "Sol is very political. He's always careful, will always do the politically advantageous thing. He'd be nice to the biggest asshole in the world if it's someone who would vote on his future." Even those inclined to a softer portrait agree that a deep political streak runs through him, one he himself appears to relish. Diane Russell remembers Snyder one time telling her he was off to "pick the brains of the wheeler-dealers."

No one who knows him questions Snyder's fondness for prizes and awards and the other "perks" of scientific superstardom. One student from the early 1970s paints him as obsessed with the Nobel Prize even back then. Says Robert Gould, "Everybody knows he'd be thrilled to pieces to get it." Edith Hendley recalls how back in his early Hopkins days Snyder had just come out of a competitive lab at NIH, and, on his own at last, felt compelled to make a name in his own right.

Which is what he did. As early as 1969, at the age of thirty-one, Snyder was awarded the Outstanding Young Scientist Award of the Maryland Academy of Sciences. The following year, he won the John Jacob Abel Award, named for the pioneering Johns Hopkins pharmacologist who discovered adrenaline. These were the first of a whole string of awards.

Once, on hearing that Snyder had won a particularly prestigious one, Julius Axelrod is said to have commented, "Oh, that's nice. Sol likes prizes." Publicly, Snyder minimizes the importance of the Nobel Prize, says he sees it as but a metaphoric yardstick for whether an experiment is worth doing in the first place, asking of it: "Will it win the Nobel Prize?"

Would he like to win it? "Oh, that would be nice," he grins.

•

From a nucleus of Edith Hendley, a single technician, and Snyder himself in the summer of 1966, Snyder's empire had, by 1970, grown to sixteen. In the cramped lab area that was

Snyder's, small cubicles had to be erected to afford some measure of private desk pace. Leslie Iversen, the British pharmacologist who'd worked with Snyder in Axelrod's lab and who would occasionally visit him in Baltimore, dubbed it "The Sollery." In building up his lab, Snyder says he consciously adopted the approach of his mentor, Axelrod: Don't get hardening of the scientific arteries by filling permanent staff positions. Rather, take on young Ph.D. students and postdocs, keep them a few years, and periodically rejuvenate with a new crop.

During these years, the field Snyder thought of as "the neurosciences" was just opening up. By no means a neatly defined discipline, it had links to psychiatry, pharmacology, biochemistry, and neurology, and people often came to him via unorthodox routes.

Joseph Coyle, for example, had been a student of French in undergraduate school, but then decided to become a psychiatrist. When, applying to medical school, he was asked whether he'd ever done research, he replied that he had—into the plays of Samuel Beckett. Somehow he got in. In his third year at Hopkins, he heard Snyder give a lecture on the brain. "It blew me away," he says, and he joined Snyder's lab for a ten-week medical school rotation. Later, he put in a two-year stint as a research associate in Julius Axelrod's lab at NIMH before returning to Hopkins and following in Snyder's footsteps as researcher and psychiatric resident.

Michael Kuhar had, at the University of Scranton in Pennsylvania, been a math and physics major. As a Hopkins Ph.D. student in biophysics, he kept hearing about a bright, young psychiatric resident named Sol Snyder who was interested in the molecular basis of mental illness. He switched graduate programs, joined Snyder in 1968, received his Ph.D. in 1970, and, after a postdoc at Yale, returned to Hopkins as assistant professor of neurosciences in 1972. "When I returned," he recalls, "Candace was here."

•

Sol Snyder has had other students over the years as bright, as imaginative, and as deeply influenced by him, but none to whom he is as indissolubly linked—through a single, overarching discovery and the stormy public wrangling that followed it—as Candace Pert. And none, it is safe to say, whose personality leaves so individual a stamp.

"At meetings you always know when Candace has entered the room," says Gavril Pasternak. Magazine and newspaper articles about her report that, for example, "her personality seems ever changing, almost mercurial," that she "is easily bored, always looking for stimulation." Another speaks of "an energy and optimism [that] sometimes threaten to explode like a Roman candle."

Both understate. One former colleague comes closer when he says, "She has such a powerful personality, when I deal with her, I sometimes feel like I've been run over."

On first meeting, she is apt to draw up close to you, cock her head, and lock her squinted eyes on yours, forging an instant, if unnerving, intimacy. As she speaks, she'll periodically heave her head from side to side for emphasis, her full locks of wavy brown hair swinging out around her like the flounced skirt of a flamenco dancer. Ideas, opinions, visions, and fantasies pour forth from her in a torrent, not always in neat and precise order, but fresh, straight from her head, without revision or artifice. One who has worked with her, Edith Hendley, says of her, "Oh, she's superb. A very bright mind. But she has an inability to edit herself. She'll blurt out things and not realize she should be more discreet. She does crazy things without thinking them out. Of course, they're very interesting things."

Candace Pert shows up at a Halloween party as a Rely tampon. She encounters a colleague's male lab assistant and smiles, "And what do you do besides standing around looking adorable?" Even her pronouncements on matters scientific are arresting: "Manic depressive psychosis is like diabetes of the dopamine receptor." Or again, the brain is "a little wet minireceiver for collective reality."

The biggest block to cancer research, Candace Pert has been heard to say, is "all these macho guys trying to beat each other out."

Men in general? "I like them in their place," she says. "Their place is the bedroom. You let them out and they start wars."

Candace Pert, née Beebe, was born on June 26, 1946, in Manhattan, and grew up in Wantaugh, Long Island. Her father, Robert Beebe, was a creative jack-of-all-trades, arranging band music, drawing cartoons, and selling radio ads. Her mother, Mildred Beebe, was a court clerk. After high school, she applied to Smith, Vassar, and the University of Michigan. Admitted to all of them, she chose a fourth school, Wheaton College, in Massachusetts. She hated it, leaving after a few months.

She transferred to Hofstra University, a local commuters' school near her Long Island home, hoping to straighten out her priorities, then head elsewhere. She got a job as secretary in the psychology department and there, in September 1965, met a student, Agu Pert. A native of Estonia, where he'd spent part of his childhood in a displaced persons camp, Agu was interested in the evolution of learning in animals. She and Agu grew close. Sometimes, when the regular animal keeper was off, they would together clean out the lab animal cages. She reports that on November 9, 1965, the day of the great northeast power blackout, she got pregnant on the floor of room 007 of the Hofstra psychology lab. She and Agu married the following March.

Once, she'd wanted to be a magazine editor. But now, though English was still her major, she was growing disillusioned. The crisis came one day when she handed in a paper, "The Greek Mind." Dynamite, she thought. C-minus, said her professor. She argued with him. It was no use. "He could say C-minus, I could say A-plus. There was no objectivity. Anybody could say anything." Disgusted, she found herself moving toward the firmer ground of Agu's science.

For graduate school, Agu Pert enrolled in Bryn Mawr College, outside Philadelphia, an elite women's college at the undergraduate level, but coeducational at the graduate. During Agu's first year, Candace—no one calls her "Candy"—stayed home, taking care of their child, Evan. But she wasn't happy, remembers Agu, and almost from the first planned to return to school.

And not in English, either, but science. Back at Hofstra, biology had been the only course she really loved. Now, with Agu, she found herself among biologists and psychologists. At home during the day, she took to reading, along with Agu's old *Playboys*, his old textbooks.

For a while she worked as a cocktail waitress at a local restaurant. One evening, she got to talking with a customer who turned out to be Bryn Mawr's assistant dean of admissions. Yes, she was thinking of going back to school, Candace told her. Where? wondered the dean. Maybe Temple, Pert replied, referring to the large private university in Philadelphia. Oh, but why not Bryn Mawr? the dean wanted to know.

It was the first time she'd considered it. Two weeks before the start of school, she applied, and was admitted.

The days of English literature courses were over. Now it was physical chemistry and psychopharmacology, all the while juggling baby sitters. Agu remembers Candace as ever enthusiastic about her studies, and recalls discussing scientific matters long into the morning. During much of this period, she slept from six to noon; it was the only time left over from classes, labs, studying, meals, and Evan.

Bryn Mawr, that most intellectually elite of women's colleges, would award Candace Pert her degree *cum laude*. ("I knew she was very gifted," says Agu.) Almost from the start it had been clear she'd be bound for graduate school. The question was, where? Agu had a military obligation to fulfill, which he planned to do at Edgewood Arsenal, an army chemical warfare research facility in Maryland.

But just where in Maryland? Neither knew. They pulled out a map, spread it out in front of them. Edgewood was about thirty-five miles from the Delaware line, twenty-five from Baltimore. Wherever Candace was going to graduate school, it had to be within commuting range of the base. Philadelphia was too far. One possibility was the University of Delaware. Another was Johns Hopkins in Baltimore. She applied to both.

Delaware accepted her. Hopkins did not.

Candace Pert attributes the rejection to "blatant [sex] discrimination. 'Tell me about your husband at Edgewood,'" she reports the Hopkins interviewer asking. How would she manage were he shipped off to Vietnam? And how, pray tell, did she expect to raise a child while attending graduate school?

Around this time, she attended the annual meeting of the Federation of American Societies of Experimental Biology, a giant conclave of perhaps twenty thousand scientists representing all the life sciences, held in Atlantic City, New Jersey. There she met a journal editor who, as Agu tells it, mentioned "this fascinating, up-and-coming guy named Sol Snyder." It was the first time she'd heard the name.

Some time later, a Johns Hopkins professor of behavioral biology, Joseph V. Brady, came up to Bryn Mawr to give a lecture. That evening, the department chairman held a party at his house, to which Brady was invited. So was Candace. At one point, the two of them danced the peabody, a popular dance from the 1920s. ("I'm probably the only person under sixty who can do it," Candace beams.) Later, she talked to Brady about her plans for grad school. She wanted to study biology and behavior, she told him—not separately, but as one; she was interested in the brain.

She remembers Brady telling her about "this weird guy Snyder," who was tackling the brain through the seemingly back-door route of pharmacology. *Snyder, again.* He was starting up a new graduate program, one quite distinct from the one that had turned her down. Why didn't she write him?

Soon, if informally, her credentials were on his desk. Three days later, near midnight, she got a call at home. It was Snyder. "You're accepted," he said. "Now apply." She came down to Baltimore one day in the spring of 1970. "This is too good to be true," she remembers thinking. She'd never seen a real research lab, at least not one like the kind at research-heavy

Hopkins. "It was very busy, very exciting." And Snyder was doing just what interested her. Until then, she'd been in something of a stew, wanting to study what it seemed could not be studied: "They said it was too complex, that there was no molecular biology of the brain." And yet this was precisely where Snyder and his group were bound.

Then, too, he seemed to take so warm an interest in his students, and in her. His friendliness and sheer human warmth —that's what struck her that first day. He even loaned her money. Underestimating the cost of the trip to Baltimore, she'd lack the cash to return. Snyder fished in his wallet and gave her twenty dollars.

She returned to Philadelphia aglow. This Sol Snyder, Agu remembers Candace telling him, was "a great guy, a very generous person."

The summer of 1970 was an idyllic one for Agu and Candace Pert. Come fall, he was to start at Edgewood, she at Hopkins. But for now, for this glorious three-month interlude, it was time out from the stresses of school, baby, and money. Agu was stationed in San Antonio, Texas, for training; but after the last few years, it was like a vacation. He was an army officer, they had some money, the world was opening up . . . and then Candace fell.

She'd been taking horseback riding lessons from an old cavalry colonel when one day she took a fall, suffered a compression fracture of the first lumbar vertebra. The hospital was crowded with wounded soldiers just back from Vietnam, some with burns over half their bodies, many of them addicted to narcotics. To the hospital doctors hers was a low-priority case. For two weeks, they fed her Demerol, an opiate drug. She loved it; it made the pain go away. Later, she felt the first signs of addictive craving, and "learned to be smart" about getting more of the drug.

At the end of the summer, the Perts moved to Maryland, Candace starting on a graduate program that would lead to her codiscovery, three years down the road, of the opiate receptor. Later she would point to her weeks in that Texas hospital, with all the mingled pain, euphoria, and drug craving, as having animated her scientific quest, giving it a personal urgency no mere intellectual curiosity could provide.

She reported to Snyder for her first day as a Hopkins graduate student. On one hand, "I thought they were lucky to have me." On the other, she was scared. Snyder put her at her ease. She was going to get her Ph.D., he assured her. They'd make her take courses, but the fewer the better, so far as he was concerned; coursework was a distraction, a necessary evil that had little to do with why she was there. She was there to do research.

In any case, he meant to get her started off right. "It's very simple," he told her. "You will apprentice to Ken Taylor. He'll teach you everything you need to know to do a histamine assay."

From her earliest days in the lab came her first scientific paper:

Young, A. B., Pert, C. B., Brown, D. G., Taylor, K. M., and Snyder, S. H. Nuclear localization of histamine in neonatal rat brain. *Science*. 173: 247–249, 1971.

Ken Taylor was an Australian and, as Pert adds, an "incredibly handsome" one at that. "When I met him, I thought I was going to faint." For months she worked beside him, learning the fundamentals of laboratory technique, performing assays, grinding up and centrifuging brains.

"I was Ken's slave," says Pert. "I didn't see Sol for months."

10.
The Opiate Receptor: "Just Get Hysterical and Do It"

EACH WEEKDAY MORNING, Candace Pert drove twenty-five miles into Baltimore from the army base at Edgewood Arsenal where she lived. "It was an army slum, very lowbrow, real ugly. The only way I got through it was to think, 'One day will be my last.' " Her husband Agu did the housework, took care of their baby, riding him over to the base childcare center on his bicycle, then had dinner ready for them in the evening.

She'd arrive at Johns Hopkins at nine in the morning, rarely return to Edgewood before seven-thirty; sometimes it was nine or ten. At the medical school, she'd park in a little alley off Madison and Wolfe streets, in a regular spot for which she paid twenty dollars a month. The area around the hospital was considered dangerous, so she'd walk from car to lab briskly. Even so, she was mugged three times in five years.

After a while the long commute became automatic, like "driving by spinal cord," she says. She began to use the time quite profitably, stopping at the 7-Eleven store, buying a big cup of coffee, planning the day's work as she drove in, visualizing each step of her experiment.

After more than a year of taking courses and gaining experience in the lab as part of her doctoral program, Candace Pert was no longer Ken Taylor's "slave." Now, she had a project of her own, one that would make her and her mentor, Sol Snyder, objects of world scientific acclaim. She was trying to find the opiate receptor.

•

For most of a century, certainly from 1905 on, the science of pharmacology rested on an assumption. Open up any textbook and there it was: A drug works by latching onto "receptors" that only it, or related compounds, fit.

The notion went back to Paul Ehrlich, discoverer of the antisyphillis drug Salvarsan, and to John Newport Langley, who in 1905 postulated a "receptive substance" on which nicotine and curare (or curari) both acted. As Langley wrote in one early paper, "I shall use the term receptive substance in describing the phenomenon of the action of nicotine and curari, although it belongs as yet to the region of theory, because its use enables the phenomena to be described in the shortest and simplest way."

Throughout the century, theoreticians of pharmacology periodically came back to the receptor concept. Yet no one had ever seen a receptor, or touched one, or even, for that matter, proved they existed. An otherwise serious pharmacology text published in 1974 offered "a friendly word of caution . . . to the inquisitive student wishing to maintain a favorable rapport with his professor. Do not request that he pass a receptor around the room neatly preserved in a specimen jar. Moreover, it would be imprudent to ask him to draw the precise chemical structure of a receptor site. At present, with few exceptions, the receptor is merely a conceptual device." It remained, in short, squarely in "the region of theory."

Transforming theory into fact became more urgent in the late 1960s, when drug abuse became a front-page issue. As many as one in four enlisted men in Vietnam were said to be addicted to heroin, while back in the States addiction was being blamed for the alarming rise in street crime. Meanwhile, even white middle-class kids were beginning to use drugs on a large scale. On June 17, 1971, in a widely publicized press conference, President Nixon launched a War on Drugs. But many basic scientists cautioned that all the addiction treatment centers in the world wouldn't crack the drug problem unless addiction itself were better understood on a molecular level.

There was one thing everyone "knew" about how heroin and the other opiates worked: They had to work on something. That something was the opiate receptor. Maybe proving its existence wouldn't mean a cure for heroin addiction, as some fevered press accounts would later assert. It would, however, represent a giant step toward understanding addiction. And from that a cure might, indeed, ultimately emerge.

Circumstantial evidence for an opiate receptor had long been building. First, there was the known existence of opiate antagonists—drugs which themselves neither produce euphoria nor mask pain yet block the action of heroin and other opiate agonists. Drugs like naloxone: give a heroin overdose victim a shot of this powerful opiate antagonist, and he'll be up and walking almost before you can pull the needle from his vein. Best explaining this near-miraculous recovery is that naloxone displaces heroin from its receptors; by occupying the presumed receptor sites, it leaves the heroin no place on which to act.

Then there is the phenomenon of saturation. Receptors, if they exist, ought to be finite in number. So that if you load them up with more and more drug you ought to get more and more pharmacological effect, until no sites remain to occupy. Sure enough, that's how most drugs, including opiates, work: a little drug, a little effect; more drug, more effect; but past a certain point, still more drug elicits no additional response.

Perhaps the most compelling clues were owed stereochemistry, the study of how the spatial arrangement of atoms influences the properties of a molecule. First, all opiates are structurally similar at the molecular level. Second, slight changes in structure can make an agonist into an antagonist, and vice versa; merely substitute an allyl group for a methyl in the agonist morphine, for example, and you get nalorphine, a potent antagonist. Both these bits of stereochemical evidence suggest a specific receptor geometry that all opiates, agonist and antagonist alike, must satisfy.

What seems to clinch the argument is the remarkable stereospecificity of the opiates: Two substances may be the same, yet not the same—the same carbons and hydrogens and nitrogens arranged in precisely the same relationship to one another save only that one is the mirror image of the other. But this subtle difference makes a difference. One version works in the body; the other is impotent. Why? For the same reason that a right-handed glove doesn't fit a left hand; though otherwise identical to its mate, no amount of turning and twisting will make the two coincide in three-dimensional space.

(This left-right terminology is no mere metaphor, but corresponds, on one level, to physical reality. A drug is said to exist in *levorotatory* or *dextrorotatory* forms, derived from Latin words for left and right, which refer to whether a solution of the substance bends polarized light to the left or right. The telltale L or D sometimes seen in the chemical formula for a drug—sometimes minus and plus signs are used instead—says which it is.)

Most substances that act on living systems do so only in their left-handed forms. Among them are the opiates. Levorphanol, for example, is a synthetic narcotic five or ten times more potent than morphine. But dextrorphan, identical except for being "right-handed," lacks all analgesic potency. *It doesn't work because it doesn't fit*—doesn't fit, presumably, the opiate receptor.

All these factors argued convincingly for the existence of something real—a molecule, a site, some special condition or configuration—that was peculiarly sensitive to opiate drugs. But in science, as in the rest of human affairs, the demand is always, prove it. And by 1971, nobody had.

•

"I hardly knew heroin from horseradish." That's how Sol Snyder describes his knowledge of opiates in 1971. But with the War on Drugs, the likely availability of grant money, and the importunings of his friend Jerome Jaffe, the "general" of Nixon's war on drugs, the problem of the opiate receptor began to intrigue him. That summer, he attended a conference on molecular pharmacology at which one of the speakers was Avram Goldstein, a Stanford University pharmacologist. Snyder took more notes on the Goldstein talk, he remembers, than all the others combined.

Goldstein had performed some experiments which, by later standards, could only be described as failures. Yet his paper, "Stereospecific and Nonspecific Interactions of the Morphine Congener Levorphanol in Subcellular Fractions of Mouse Brain," which appeared in the *Proceedings of the National Academy of Science* in 1971, would later be seen as the parent of all future work, his experimental strategy *the* model to follow.

You want to go after the opiate receptor? Goldstein asked. Here's how: First, he pointed out, a drug bound to a piece of tissue need not be bound to receptors in that tissue, because there are many ways for one substance to bind to another, including ionic bonds, hydrogen bonds, hydrophobic forces, and the like that have nothing to do with how the drug really works. Such binding is nonspecific. Pour a drug onto tissue and some of it sticks; the task remains to distinguish what clings to the tissue in these meaningless ways from what fits its receptors in a pharmacologically meaningful way.

The first element of Goldstein's strategy was to overload a piece of tissue, like mouse brain, with great gobs of an opiate drug like levorphanol so that every receptor site is, presumably, filled. What if you then added radioactive levorphanol? Presumably, since the receptor sites are already occupied, the radioactive opiate would find nothing to which to bind. Thus, a radioactivity count of zero.

Of course, that's not what happens at all. Take the radioactivity count and it's pretty high. Although all receptor sites are occupied, other ways remain for the drug to cling to the tissue. However much does is a measure of nonspecific binding, the kind which results from every kind of molecular interaction but that to receptors.

Next experiment in Goldstein's plan: Take dextrorphan, the pharmacologically inactive form of levorphanol, and soak a piece of brain tissue in it. The tissue is left clogged with dextrorphan everywhere except for the receptors—because the "right-handed" dextrorphan can't fit these "left-handed" sites. Now, once again add radioactive levorphanol. The would-be receptors, unoccupied by dextrorphan, are free to fill with hot levorphanol. So that the radioactivity you now count would be a measure of receptor binding, except for one factor: You don't know how much of what clings to the tissue actually represents nonspecific binding.

However, you really do know—from the first experiment. Subtract the radioactivity recorded in the first experiment from that in the second and you've got a measure of stereospecific receptor binding. A solid enough number and you'd be on your way to demonstrating the opiate receptor.

Goldstein carried out his plan. With one mouse brain, for example, the second experiment gave 2,521 counts of radioactivity per minute, the first 2,298, for a difference of 223. All told, using eight mouse brains, about two percent of the total binding was stereospecific—evidence for receptors that was feeble at best.

Later, it turned out that the binding Goldstein had reported was not to the opiate receptor at all but to another substance which could also tell left from right, as it were. But even at face value the result inspired little confidence. It was too muddied, too iffy. If receptor binding counted so heavily in the working of drugs, as all evidence suggested it did, the crucial experiment ought to fairly shout out the news.

Still, Goldstein's experimental strategy held promise. Snyder remembers wondering how it could be refined so as to get clearer results, scribbling his own ideas all over Goldstein's paper. All that was needed, to hear him tell it, was the right student to carry out his experimental strategy. Sometime in late 1971 or early 1972, he assigned Candace Pert to the job.

•

In relating the events that led her to work on the opiate receptor, Pert lays stress on quite different factors, including her horseback riding accident in Texas, her painful hospital stay, and the prescribed diet of opiate painkillers that almost addicted her.

While still in the hospital, she'd phoned Snyder, asking what she should read for a head start on her graduate studies. Oh, just take it easy, he advised, but if she were really serious, why didn't she look at *Principles of Drug Action*, by Avram Goldstein, Lewis Aronow, and S. M. Kalman. Its first fifteen pages were devoted to the receptor concept. "Everybody knew there were receptors," she says. Only no one had proved as much.

Some time after joining the lab, she says, came a dinner at Sol Snyder's house. The dinner was by way of welcome to her and new faculty member Pedro Cuatrecasas, and their spouses. Pert told of her drug experience at the Texas hospital, Snyder and Cuatrecasas both listening with what she read as "morbid interest." She and Agu, from whom she is now divorced, both remember the opiate receptor coming up that evening as a scientific problem she might consider tackling.

While getting her required courses out of the way, Pert also did a series of lab rotations. One of them, for five months, was in Cuatrecasas's laboratory. "If anyone in the department will win the Nobel Prize, it will be Pedro," Pert remembers hearing. He was brilliant and creative, and everyone knew it.

Cuatrecasas was working not on the nervous system but on insulin; the hormone secreted by the pancreas that helps control blood sugar levels, lack of which causes diabetes. Receptors, the assumption went, figured not just in the nervous system but wherever drugs and hormones act. On what did insulin act? Presumably on insulin receptors. And Cuatrecasas had pioneered techniques to demonstrate them.

Cuatrecasas remembers Pert as attentive and enthusiastic but not among the most scientifically original grad students to come through his lab. At the bench, she was sloppy, with small patience for experimental drudgery. Moreover, at first she was "not rigorous in her thinking or her interpretations." (In response to such perceived deficiencies, there'd been some move to dismiss her from the graduate program. It was Snyder, by all accounts, who blocked it.)

On the other hand, says Cuatrecasas, Pert was quick to learn and, in her exuberance, "a joy to work with." To criticism of her lab work and appeals for clearer, closer thinking, she was responsive. She listened, and in time, the lessons took. She grew "more careful, less sloppy—no question about that."

Pert learned much in Cuatrecasas's lab. But most of all, to hear her tell it, "I learned how to do binding assays—everything I needed to know to find the opiate receptor."

That Pert would work on the opiate receptor was no foregone conclusion. It was a tough problem for a new graduate student. Snyder had joked, "It's easy—just like the insulin receptor." But it wasn't just like the insulin receptor. In fact, says Pedro Cuatrecasas, "at that time to look for a receptor for a drug not found in the body was a little bold." The insulin receptor, after all, was there because of insulin. What was the opiate receptor for? Heroin?!?

Snyder at first had Pert working on choline uptake, more of a meat-and-potatoes problem that was sure to leave her with an easy Ph.D. But Pert had scant interest in it, and hankered for the more ambitious opiate receptor problem. Finally, after a time during which Snyder felt she gave choline uptake only lackluster attention, he proposed to her the opiate receptor instead. Her Ph.D. topic. Her baby.

Early in 1972, she set to work. At the very outset, Snyder gave her Goldstein's seminal 1971 paper to read. All through the spring and summer she tried variations on the theme of the Goldstein strategy, dropping it from time to time, then picking it up again for another shot. She got nowhere. Stereospecific binding? Zero. Zilch. *Nada.*

Soon Snyder grew fidgety. Everything he'd learned from Julius Axelrod predisposed him to easy answers and doable problems, and the opiate receptor was beginning to seem distinctly undoable, and certainly not easy. The last thing you wanted was to settle into a grim lockstep toward some scientific Holy Grail. You could get bogged down for years that way. Besides, he had a Ph.D. student relying on his guidance; it wasn't fair to her. The opiate receptor showed little sign of immediate success. He was inclined to drop it.

She wasn't. In that respect, she says, her work on the opiate receptor was not in the Axelrod tradition. But she knew, just *knew*, that there was an opiate receptor, and that she could find it. "It was my obsession," she says. "It was all I wanted to do."

•

Success, when it came, was built up from numerous small successes, most of which appeared, in the final paper by Pert and Snyder, too fleetingly and too innocuously to give much clue to their significance, like this one, in the second paragraph, camouflaged in a thicket of methodological detail:

> Samples were cooled to 4 degrees Celsius, filtered through Whatman glass fiber circles (GF-B) and the filters were washed under vacuum with two 8-ml portions of ice-cold tris buffer.

Pert and Snyder were describing how they treated the samples of ground up rat brain they had allowed to incubate with radioactive drug. It was the solution to one problem that had dogged Goldstein, whose specific binding had been so blurred by indiscriminate, pharmacologically meaningless binding: The filter held brain tissue. By washing it under vacuum, unwanted "dirt"—radioactive drug loosely bound to everything but receptors—was carried away. Presumably, the receptor-bound drug remained behind.

The technique Pert and Snyder described was similar to that reported in another paper, appearing two years before, in the *Proceedings of the National Academy of Sciences*: "3 ml of ice-cold KRB-0.1% albumin is added to the cells, which are immediately filtered and washed with another 10 ml, under reduced pressure, on cellulose acetate EAWP Millipore filters." Here it wasn't brain tissue being filtered but fat cells, not the opiate receptor being sought but the insulin receptor, not Pert and Snyder who were looking for it, but Pedro Cuatrecasas.

Cuatrecasas's rapid filter-and-wash method may be loosely likened to a properly exposed photograph. Exposed for just the right length of time, a photographic negative records every detail—the blackest blacks, the whitest whites, the grays in between. But overexpose it and the final picture is washed out, all detail lost. The filter-and-wash technique limits the tissue's exposure to radioactive drug; the "detail" retained is receptor binding unmuddied by nonspecific binding.

In the end, the filter-and-wash method was crucial. Yet while solving one problem, it potentially aggravated another. Pert was using radioactive drug at concentrations far lower than those used by Goldstein, and then washing most of it away. The amount of drug left bound to the receptor thus risked becoming too little to count at all. The need, therefore, was for not just a "warm" drug but one that was truly "hot," one with high specific radioactivity.

The first drug Pert tried was dihydromorphine, a thousand times hotter than anything Goldstein had used. It might have worked, they learned later, except that dihydromorphine degrades under normal lab lighting conditions. That they didn't know till later.

Failure, again and again. But Pert kept at it, trying different drugs, changing temperatures and incubation times, refining the washing technique. Nothing worked.

Then one day, on September 22, 1972, it did work.

•

In the dedication to her doctoral dissertation, Candace Pert wrote: "For Agu, who has given me love, Evan, encouragement, and naloxone." It was naloxone that unlocked the door to the opiate receptor.

While Candace worked in Baltimore, husband Agu was an

army chemical warfare researcher at Edgewood Arsenal. Into the brains of monkeys he would insert hollow stainless steel needles, through which morphine or other drugs could be injected. The level of analgesia induced could be measured by how much "standard" pain—time on a hot plate, for example—it suppressed. Selecting different brain regions, Agu sought to pinpoint just where morphine acted. Then, as a check, once he'd found such an area, he'd inject other opiate agonists through the same needle, anticipating continued pain relief; or else he'd inject an opiate antagonist, in which case the analgesia ought to be terminated. One drug Agu used was naloxone, a powerful opiate antagonist.

In her until now futile quest for the opiate receptor, Pert had used radioactive agonists. But what about antagonists? A long paper by the English pharmacologist W. D. M. Paton had outlined a theory to explain why antagonists worked differently from agonists. Maybe, it had given her cause to think, a radioactive antagonist would compete for receptor sites more fiercely than the agonists she'd used until then. An antagonist, for example, like naloxone; Agu had lots of it at Edgewood.

Agu Pert's straight blond hair, bushy blond mustache, and smiling, slitted eyes make him look like he ought to be wearing a lumberman's jacket and appearing in Camel cigarette ads. He is, in many respects, Candace Pert's polar opposite: quiet, sometimes so quiet you can barely hear him, with a hint of the self-deprecatory. He is divorced from Candace now, but he supports in almost every particular her recollection of the events leading up to the discovery of the opiate receptor. "The naloxone idea itself was clearly hers," says Agu evenly. "I can vouch for that. It was generated by reading that article."

Agu could supply all the cold naloxone Candace wished. But tritiated naloxone, hot naloxone, had to be custom made. All through the summer, she'd been having New England Nuclear prepare various radioactive drugs for her experiments. But by now, Snyder was leery about sinking more money, or more time, into the project. Dare she go ahead and, on her own, send off a batch of naloxone to New England Nuclear? "She agonized over it," says Agu, "then finally just went ahead and did it."

She rolled up a few milligrams of the powder into a wad of paper and mailed it off. She didn't, she says, tell Snyder. "I remember feeling slightly guilty about it."

Sol Snyder remembers it differently. He describes the decision to order tritiated naloxone as a joint one, inquiries to New England Nuclear being at his direction, and the order for it going out with his knowledge and blessing.

Whatever the circumstances under which the cold naloxone left Hopkins, it came back a month or so later, in liquid form, highly radioactive, and in lead-lined casing. Pert reported to the Hopkins Radiation Control department to pick it up, and there performed the initial steps in its purification.

Then it was back to the lab to try the binding assay. It was September 22, 1972.

The next day Pert marched into Snyder's office and handed him the data from the previous day's work. "Look at this," she said. "You're not going to believe it."

For a moment, she remembers, he sat quietly looking at the numbers. Then he let out a whoop of joy. "Fuck! Fuck! Fuck!" he exclaimed. He was up, out of his chair, "cursing around the room."

Once he'd quieted down, they went over the experiment in detail. This preliminary result was just what they'd been looking for: clear stereospecific binding. They had only to more methodically confirm it. To do that, Pert needed help, and now Snyder was ready to pull out the stops to furnish it. He approached master technician Adele Snowman. "Why don't you give Candace a hand for a couple of months?" he asked her.

Snowman would get to the lab at five-thirty in the morning, leave early in the afternoon. Pert arrived at nine, stayed until eight. All day, they did binding assays. Today, Snowman uses a commercial instrument that enables her to read as many as eighteen hundred scintillation vials a day; earlier generations of the instrument let her do five hundred. Back in 1972, a fair day's work was a couple of dozen. Even so, she and Pert churned out reams of data. "Candace worked an incredible number of hours," recalls Agu. All the while Snyder was urging her on: "Hurry up, we'll be scooped!"

Pert had performed the first successful experiment in late September. Their joint paper—Snyder remembers the two of them writing the first draft in two hours—reached *Science* on December 1 and later, in revised form, on January 15. "Pharmacological evidence for the existence of a specific opiate receptor is compelling, but heretofore it has not been directly demonstrated biochemically," it began. "We report here a direct demonstration of opiate receptor binding, its localization in nervous tissue, and a close parallel between the pharmacologic potency of opiates and their affinity for receptor binding."

Their paper satisfied all the standards of proof Avram Goldstein had established two years before. The central experiment was, in essence, Goldstein's. To recap: If minced rat brain is soaked with the agonist levorphanol, thus occupying all receptor sites, any added radioactive naloxone clinging to it must represent nonspecific binding. So the corresponding radioactivity count must be subtracted from that recorded when no levorphanol occupies the receptor sites. In a typical experiment, they'd get eight hundred counts per minute with levorphanol present, two thousand when it wasn't. Goldstein had been encouraged with two percent specific binding? They got sixty percent!

But this was just the foundation of the edifice of evidence they built up into proof of the opiate receptor. They found, for example, that the extent to which opiates other than levorphanol interfered with naloxone binding correlated almost perfectly with their known pharmacological potencies; while it took only low doses of morphine to interfere with naloxone binding (to a given and arbitrary degree), a much higher concentration of codeine was needed to achieve the same effect. In short, the stronger the drug the more fiercely it competed with naloxone for the receptor—exactly as predicted by receptor theory.

As a double check, Pert tried every nonopiate drug she could think of—serotonin, atropine, caffeine, histamine, and many more—to see if they'd compete with naloxone for receptor sites. It would have thrown a monkey wrench into the works had they done so. They didn't.

Finally, they repeated their experiments, not on the brain as a whole this time but on various parts of the brain. The corpus striatum, they reported in that first paper, showed the most receptor binding. In later work they pinpointed the brain's limbic system, known to play a role in the perception of pain, as especially rich in opiate receptors.

Their big paper, "Opiate Receptor: Demonstration in Nervous Tissue," with Pert the lead author, appeared in the March 9, 1973, issue of *Science*. A few days before, the news went out:

> BALTIMORE (UPI)—Two Johns Hopkins Medical School researchers have been credited with a major breakthrough that could lead to a better treatment for narcotics addiction, the National Institute of Mental Health has announced.
>
> A professor and a doctoral candidate discovered for the first time the areas of the brain believed to transmit the effects of narcotics—including euphoria, relief from pain, and the causes of addiction.
>
> "What we have is not a cure for heroin addiction, but something that may lead us to a faster cure than we had hoped," said Candace B. Pert, a graduate student in pharmacology. Ms. Pert and Dr. Solomon H. Snyder revealed their discovery of the brain receptor sites after a year of research supported by NIMH. . . .

The opiate receptor was big news, and the press conference at which it was announced made it bigger yet. The War on Drugs had made addiction-related research a political issue that reached all the way to the White House. With Vietnam and Watergate sapping the country's morale, the opiate receptor was seen as something positive and dramatic to offer the public. NIMH, which had supported the research, decided to play it up big, and a decision was made to hold a large press conference, orchestrated by the Johns Hopkins public relations department, in Baltimore.

"All of a sudden," recalls Gavril Pasternak, a veteran of the lab from that period, "the lights of the world were on us. There were ten thousand TV cameras. There was *Newsweek*. There was *U.S. News*." The wire services were there. So were the *Washington Post* and the *New York Times*.

And facing the lights and cameras, introduced as co-discoverer of the opiate receptor, was Candace Pert, age twenty-six, her long brown hair billowing down over her white lab coat, sharing the stage with her mentor. "Dr. Pert, can you tell us . . . ?"

Later that year, Pert presented details of the receptor work at a meeting of the International Narcotic Research Club in Chapel Hill, North Carolina. All the big boys of the opiate field were there, like Albert Herz of the Max Planck Institute for Psychiatry in Munich, Germany, and Vincent Dole of Rockefeller University, and Hans Kosterlitz, the seventy-year-old German-born scientist who had fled the Nazis before the war and made Aberdeen, Scotland, into a world capital of opiates research. Herz and Kosterlitz hugged her, remembers Pert. Kosterlitz took her to dinner with some drug company executives. It was a heady time.

In the glow of good will that washed over her at the Chapel Hill meeting, Candace Pert saw a model for science at its best. Herz and Kosterlitz, she felt, were genuinely happy for her. Science didn't *have* to be dog-eat-dog. "The whole competition shit—that's not what science is about."

But back at the lab, there was plenty of competition: Gavril Pasternak was getting a piece of the now rapidly expanding opiate receptor pie. A native of Brooklyn, Pasternak's entire higher education, spanning fourteen years from bachelor's degree in chemistry all the way through medical internship and neurology residency, would be spent at Johns Hopkins; now he was picking up a Ph.D. to go with his M.D., and almost from the moment he arrived in the lab, he and Pert clashed.

Their rivalry came to a head following the curious and quite accidental discovery that sodium ions enhanced the binding of opiate antagonists, while inhibiting that of agonists —affording a handy *in vitro*, or in-the-test-tube, means for discriminating between the two classes of drug. Pert, Pasternak, and Adele Snowman all had a hand in it, but Pert felt Pasternak had infringed on her scientific turf. Called in to referee, Snyder awarded Pert first authorship of the big paper. To Pasternak he granted custody of work that ultimately came to little. Pasternak felt cheated.

After that, the two of them scarcely talked. It was Snyder and Wurtman all over again, a scientific sibling rivalry.

The subsequent hooplah over the opiate receptor left Pasternak feeling ignored. "Obviously, I was quite jealous," he admits—and it came out. Snyder remembers spending hours trying to quell shouting matches between him and Pert. "I don't envy Sol putting up with Candace and me," says Pasternak. "I was as bad as Candace. We were like two kids fighting." It was a competition for Snyder's attention, remembers Michael Kuhar, who'd returned to Hopkins from a Yale postdoc to find them already at one another's throats.

Adele Snowman, who got along better with Pasternak, watched the rivalry unfold. "They wouldn't listen to each other. They were each right in their own ways, but their personalities didn't mix." Neither did their scientific styles. "I'll obsess about something," says Pasternak, "do it ten times. She'll do it once and publish it." Pert, in her turn, simply writes off Pasternak as a poor scientist. How so? "Too rigid," she declares.

The enmity between the two has eased little over the years and even more than a decade later neither could muster much nice to say about the other. Pasternak pictures Pert as "settling into her place now"—a place he says, voice heavy with implication, much diminished from what it was during the glory days of 1973. While conceding she's extremely bright and crediting her with "flashes of brilliance," he portrays her as indulging in scientific tangents that lead nowhere, with sloppiness in the lab as well as of the intellect.

"Probably even worse," he laughs, is Pert's assessment of him. But while Pert does reciprocate, it is not in equal measure and, she insists, "I bear him no rancor."

It was painful for them both, admits Pasternak, reflecting back on their time together in the lab. "And Sol was caught in the crossfire."

•

The opiate receptor had launched a revolution and Snyder had fired its opening shots. Suddenly, as a friend recalls, he was "this boy wonder going up a million miles an hour." Eager young grad students and postdocs flocked to join his lab. Grant money poured in. A whole string of awards came his way. The John Gaddum Memorial Award, from the British Pharmacological Society, in 1974. The Efron Award, from the American College of Neuropsychopharmacology, the same year. Snyder was named Cambridge University's Sir Henry Dale Centennial Lecturer in 1975, the University of Wisconsin's Rennebohm Lecturer in 1976, and received the Van Giesen Award from Columbia University in 1977. Many others followed. In 1978, he was named president of the six-thousand-member Society for Neuroscience.

Hopkins early recognized it had a scientific superstar on its hands. Even before the opiate receptor discovery, when he was thirty-one, Snyder had been made a full professor—the youngest in Hopkins' history—and in 1977 he was named Distinguished Service Professor. Then, three years later, Johns Hopkins announced the creation of its first new basic science department in twenty years, the Department of Neuroscience. Named to head it? Solomon H. Snyder, who would soon move upstairs from his old, cramped lab on the third floor of the Wood Basic Science Building to spacious, oak-accented quarters on the eighth floor replete with beige carpeting, recessed lighting, and framed works of art.

The opiate receptor had opened up whole new territory. It was not only new insights into the nature of addiction that it offered, nor even its potential as a tool to study addiction. Rather, it represented a powerful new technology—receptor technology—for probing the workings of the body as a whole, the nervous system, and the mind. Does Valium, say, work as a sedative? Does caffeine keep you up at night? Then there must be receptors upon which those drugs act, receptors that could be studied in the same way as the opiate receptor. The same went for any of a lengthening list of brain neurotransmitters, each of which, presumably, also had its own receptors.

There were clinical applications, too; by 1979, for example, Snyder's lab would use receptor technology as the basis for a simple new means of custom tailoring the dosage of antischizophrenic drugs to individual patients. In private industry too, opportunity beckoned. Where once drugs had to be screened on living animals, now you could perform crucial assays right in the test tube, a single rat brain conceivably enough for thousands of experiments. Drug development, Snyder would estimate, could be speeded up a hundredfold.

During the mid-1970s, the lab astir with activity, a new crop of grad students and postdocs began following up some of these leads. Candace Pert recalls how old hands like her, with the haughty air of those who have discovered a restaurant that later becomes fashionable, "looked down their noses at the new people." She'd joined Snyder because his scientific interests coincided with her own, while the newcomers, some of them, had "found their way to Sol because they were looking for an established scientist."

David Bylund, who, for two years beginning in 1975, was a Snyder postdoc, remembers how in those days, everybody had his or her own receptor. His was the beta-receptor, one of the two through which adrenaline acts. Someone else had the alpha-receptor, another the gamma-aminobutyric acid (GABA) receptor, and so on. Each posed methodological problems that in retrospect, according to Bylund, should have been trivial. Somehow they never were. Temperatures, buffers, concentrations, filtration conditions: Bylund needed almost four months to work out the details for his paper with Snyder, "Beta-Receptor Binding in Membrane Preparations from Mammalian Brain." Appearing in a 1976 issue of *Molecular Pharmacology*, it was Bylund's fifth published paper, Snyder's 274th.

"Like Florence in the Renaissance." That's how Candace Pert has described the field opened up by the opiate receptor. The discovery thrust Snyder's lab into the scientific big leagues. A flurry of other discoveries kept it there, solidifying its reputation. Students poured into Snyder's lab and then, once their time with him was up, descended upon the scientific world, taking something of his research style, his whole approach to science, with them.

•

You could scarcely work in Sol Snyder's lab without feeling the almost palpable presence of Julius Axelrod. "There is a look in people's eyes when they respect someone," says Gavril Pasternak, "and you could tell that Sol held Julie in the highest regard."

Diane Russell remembers how, back in the late 1960s, Snyder talked with him almost weekly, and how what Axelrod would think of something always counted heavily.

Robert Gould, from a much later period, in the early 1980s, recalls how Snyder, by then fifteen years out of Axelrod's lab, would sometimes ask out loud, " 'What would Julie do?' It was like calling on the name of God when you need inspiration."

What *would* Julie do? He wouldn't waste time on the trivial or the impossible. He'd keep it simple. He'd do it fast. He'd take a flier. And that's pretty much what Snyder's students took from him, too.

Reviewing her career in the *Annals of the New York Academy of Sciences*, Diane Russell credited Snyder with introducing her to the polyamines, her life's work, and with teaching her to critically interpret scientific papers. But other, less tangible lessons stuck with her as well. "We all have about the same number of hours of the day to do things," she remembers Snyder saying. Why waste them? Therefore, choose your problem with exquisite care, distinguishing those that are merely interesting from those that are important as well.

Yet you never wanted to get hung up on a problem which, important or not, left you huffing and puffing away with no great likelihood of success. Snyder had almost a sixth sense for scientific questions apt to leave him batting his head against the wall—and he avoided them.

The average scientist, explains Robert Goodman, keeps a particular scientific goal always before him, like finding a gene or purifying a particular enzyme. He'll be prepared to expend much time and much effort before getting the answer, or giving up. "That's definitely not the Sol Snyder approach. He doesn't like to get bogged down. He doesn't want to spend a year on a problem and then say it doesn't work."

Brain researchers David Hubel and Torsten Wiesel, to pick a classic example of the opposite approach, spent most of the 1960s and 1970s exploring the architecture of the brain's visual cortex—methodically, step by step, forever deepening their understanding, trying one thing, then another, refining, embellishing. Ultimately, after twenty years, they'd drawn a richly detailed portrait of how the brain interprets visual imagery, which in 1981 won them the Nobel Prize. Theirs was a scientific style quite different from Snyder's. Not better, not worse, just different.

"Sol's approach," says Goodman, "is never to establish an end point ahead of time. He goes where things take him. He'll revise plans, make a lot of whatever's there." He'll take a flier.

Robert Gould tells how Snyder was always encouraging them to try off-the-wall things, suggesting this or that approach that, often enough, seemed outrageous. Yet then again, why not? "Sol's attitude is: Do the experiment. Find out." Many a scientists sits in the office and thinks it all out beforehand, then does the experiment so thoroughly that, as Gould says, "you never have to do it again. Snyder would rather do five rough experiments."

Snyder's reasoning was so obvious it scarcely needed stating, yet did: "I'd rather do an experiment in three hours than three days or three months," he says. "That's just common sense. You'd be surprised how many people don't have any." Around the lab, he'd recite Julius Axelrod's dictum to the effect that an experiment is worth doing only so long as it's easy to do. That is, an experiment demanding elaborate preparation and riddled with obvious pitfalls may not be worth tackling at all—not so long as you've got a dozen equally important ideas that are a snap to try.

For Candace Pert, what she calls Snyder's "pragmatic, handyman approach" to science was a revelation. He was always sidestepping the gray muck of experimental tedium, always scaling the heady scientific heights, reaching for the fundamental, more exciting problems that sneered at routine. He went right for what he wanted: Need a new technique just appearing in the scientific literature? Don't spend days in the library poring over journals trying to figure it out; just call up its originator and get the details directly. Spy a striking new tack to take with a problem? Don't worry about elaborate scientific controls for now: "Just get hysterical and do it," as she describes his style. "Go for the instant gratification."

Edith Hendley, too, recalls her years with Snyder as "the best thing that ever happened to me. My whole approach to research was forever influenced by him—you know, not to get bogged down in perfecting the details, but to look at the big picture." Many scientists choose to perfect one thing before going on to the next. Not Snyder. He'd stay with it just so long as his curiosity allowed, and no longer. "I learned from him," says Hendley, "to just charge ahead."

For Gavril Pasternak, Snyder is a scientific Daniel Boone intent on blazing a trail across the continent, not on leisurely surveying every stone and plant within a small area. There's much detail you never see that way, but you do get to where you're going first. Snyder scorns elaborate follow-up; rather, says Pasternak, he'll flit from one subject to another, pursuing the next hot lead, letting scientists of more enduring patience fill the gaps left by his trailblazing.

Which happens to be one of the things about Snyder that scientists who don't like him don't like.

•

An incident in Diane Russell's relationship with Snyder perplexes her still. It was back in the late 1960s, and for two years she'd been working on polyamines at a National Cancer Institute research center in Baltimore, a position Snyder had helped her get. A conference was coming up and she and Snyder had informally agreed to present a joint paper. Snyder was to write it, including within it their work together, as well as her more recent work since, much of it yet unpublished.

When Russell saw a draft of it, it seemed to her that "everything I'd been doing those last two years was Sol's." The paper did not claim her work was the product of his scientific leadership. But he was senior author, and without specific assurances to the contrary that's how it looked, at least to her. Deciding to take a stand, she made an appointment to see him.

The day of their meeting, she drove across town from the cancer center, near Johns Hopkins's Homewood campus in north Baltimore, to the Hopkins medical complex in east Baltimore, nervous the entire way.

Yet when she finally confronted him, no argument ensued, no hard feelings. "OK, we'll lop it off," Russell remembers Snyder agreeing easily. Their original joint work on polyamines would appear in a joint article, they resolved, while her most recent work would appear in a separate paper, labeled "Discussion," which she'd present herself.

The episode, Russell says, may reveal more about her own development as a scientific professional than it does about Snyder. "It was a turning point, to recognize my responsibility to my work," she says. Snyder's acquiescence suggests he'd merely let his eagerness to tie up the whole subject in a neat intellectual bundle run away with him. "That he backed down suggests it was not intentional."

Or so she chooses to interpret it today. "That way we can be friends," she says. "Otherwise I'd have to bear him ill feeling." Was Snyder implicitly taking credit for her work? She doesn't think so. "And if he was," she adds, "I don't want to know about it."

The incident, with all its ambiguity, highlights the curious ambivalence with which many of Snyder's colleagues view him. In one breath they laud his creativity and intelligence; in the next they decry what they see as his stage-managed scientific success. In the driving urgency that informs his scientific style and stirs the juices of his students, meanwhile, they find strains of unchecked ambition.

Edith Hendley, otherwise a Snyder fan, admits she's heard her mentor described as overly ambitious, even ruthless. Once, the rumor circulated that Snyder had scooped a young NIH scientist and "grabbed her finding. It made me sick to hear." Was it true? She doesn't know.

A researcher of distinctly minor reputation at a Washington, D.C.-area medical school says she's heard colleagues say, " 'Don't send [a paper for review] to Sol. He'll steal it." By *steal* she means he would, guided by its findings, put someone in his own lab to work on the problem; that he then might publish the results in a journal with a short lead time, thus claiming priority for the discovery. Examples? She can't supply any. Pressed, she still offers none. "But that kind of reputation doesn't develop," she says ominously, "unless its justified.

"I don't trust Sol," she goes on. "Many people don't trust Sol. He's the single hungriest person I've ever known. He'd do anything to get the credit for a discovery."

Whether or not any of this is true bears not alone on Sol Snyder's character, nor even on just his personal reputation as a scientist. For to whatever extent Snyder is emblematic of Big Science today, it bears on science's essential workings. And it casts light, too, on the controversy ignited later, in 1978, when the Lasker Award committee assigned credit for the opiate receptor to him and not to Candace Pert.

According to Gavril Pasternak, a natural storyteller who narrates his tales of scientific infighting with obvious delight, it's to the discovery of the opiate receptor that ambivalence toward Snyder can be traced. "You have to keep in mind what happened in 1973."

At the time, he explains, most of the old hands in opiates research were members of "The Club." It is called something more professional-sounding today, but until the mid-1970s there really was something called the International Narcotic Research Club. Snyder was not a member. He might send his students to its meetings, but never showed up himself. After all, he'd never before even worked in opiates.

Yet it was this same Snyder who, in 1973, with Pert, discovered the opiate receptor. In six months. "So you get this brash young guy who sets the whole field on its ear, making all of them look foolish." For them, opiates were their life, while "for Sol it was one small part of his work. . . . He not only made them look foolish, he did it part-time."

Within this circle, he asserts, Snyder and his people remain outsiders even today. "The world of opiates," says Pasternak, shaking his head, "is a vicious, dog-eat-dog world."

And maybe not just the world of opiates, to judge from the depths of bitterness Snyder's name sometimes stirs outside it. Many neuroscientists and pharmacologists, perhaps most, regard Snyder with infinite respect, unbounded admiration. Others, though, say of him that he'll enter a new field only to ignore what the oldtimers have done, failing to properly credit their work. It's not illegal, says one, just shady. Another scientific competitor charges Snyder with "unethical practices that are close to stealing."

Snyder's name will come up in a list of eminent scientists and the speaker will interrupt himself to say, "He's brilliant, you know." Yet this same individual will say Snyder has achieved his success by blowing his own horn and through other means none too ethical. "Snyder is very intelligent," says a former coworker from their days in Axelrod's lab. "But his ego is very large. He wants to do everything; to be visible. He pushes out great quantities of information."

Simple envy? That's how Robert Goodman, for one, sees it. Snyder's critics, he says, just don't understand how he works—and fail to achieve anything like his results. While a respectable lifetime's output for a scientist might be fifty or a hundred published papers, Snyder had four hundred by his fortieth birthday. "They're jealous," says Goodman.

But of this great Niagara of papers, critics point out, too many are simply wrong. Evidence, for example, for a single opiate receptor at which drugs may bind at distinct sites? Disproved. A finding that Valium and the other benzodiazepines work at the same receptor sites as the neurotransmitter glycine? Also wrong.

Neither Snyder nor his admirers deny that the hard-charging style of his lab—churning out ideas, doing experiments, guessing, trying, speculating, spitting out the papers—sometimes leads to mistakes. And of course, Snyder regrets them. On the multi-sited opiate receptor, for example, Snyder says, "we had our hands over our eyes. We squandered a year and a half before we got the story straight. In science, you're always trying to make sense of complicated data, and sometimes you tend to disregard things that don't fit your theory."

But the errors, when they come, are confined to interpretation, Robert Gould takes care to note; no one raises the specter of outright data-fudging of the kind that's made headlines in recent years. In fact, says Gould, "because Sol sometimes makes such outrageous leaps of faith, you lean over backward to supply him with accurate data" on which to base them.

Snyder, explains Gould, feels an invalid interpretation doesn't mean you're stupid, or a bad scientist, or anything of the sort. After all, make conservative enough claims and you're never going to be wrong. "You can be nine million percent sure and guarantee you make no mistakes," says Snyder. "But you also won't discover anything."

Better to pursue a more ambitious problem which, though apt to send you up blind alleys and invite shaky speculation, promises a bigger payoff when it does pan out. "Aim high," Snyder counsels. "A student will say, 'It's good science, isn't it?' But if I'm sitting there falling asleep while he's telling me the best it can work out, well. . . . So I'll say, 'Yes, but it's boring. I think we can do something more exciting.'

"You can't be a chicken shit. So you make a few mistakes; the world won't come to an end." But it's just this devil-may-care detail-glossing that troubles his critics.

With Snyder a powerful figure in the neurosciences today, most who fault him do so only off the record. One who does publicly, however, is Theodore W. Rall, professor of pharmacology at the Universtiy of Virginia School of Medicine. Rall applauds Snyder for his creativity, his intelligence, and his unwillingness to let gaps in his knowledge stand in the way of making discoveries. But he feels, too, that Snyder's papers tend to "sweep things under the carpet to make the correlation sound better. . . . There's that attitude of 'Let's always put the best foot forward.' He's always selling something."

Yes, one is apt to tolerate such failings in someone as scientifically adventurous as Snyder, Rall admits. Tolerate it, that is, until you pick up your local newspaper, as Rall did one day in 1981, and see heralded as a Snyder discovery that caffeine and the other methylxanthines work by blockading adenosine receptors. There it was, the news spread across the country by wire service, "trumpeted about, without Snyder's saying 'boo' about anything that had happened before 1981." Yet in the sixth edition of Goodman and Gilman, the standard pharmacology text, says Rall, he had himself pointed to the blockade of receptors for adenosine as the likely mechanism by which the methylxanthines act. And the experimental evidence, some of which he, Rall, had furnished, went back years.

Rall also scolds Snyder for failing to properly credit the work of others in a paper on the enzyme enkephalin convertase; for erroneously reporting that Valium worked through its effects on the glycine receptor, when it seems rather to work through the receptor for a neurotransmitter called GABA; and so on.

What irks Rall especially is that when Snyder does err he seems content to simply stride on, oblivious, trumpeting some more recent discovery or another. "Don't bother me with the details," is the attitude he ascribes to Snyder. "Take the evidence and advertise it. Take it around to meetings. Get lots of papers. Get lots of grants.

"I don't condemn him for being wrong," Rall goes on, "only for being wrong and not picking up that it's wrong. You should try to verify or shoot down your results. Don't leave it to someone else. That's too much like a snake oil salesman. It's unattractive and, well, unscrupulous."

Nor, he adds, can Snyder's scientific recklessness be mistaken for the almost naive ebullience of a Julius Axelrod. " 'Gee, look what I found,' is Julie. He's like a kid with Christmas candy." Whereas Snyder, to his taste, is a glory-grabbing victim of what he calls "Nobelitis."

And how does Snyder himself reply to Rall's charges? As to the caffeine, Snyder says that yes, Rall and a colleague had indeed found, some years before, that adenosine's effects on cyclic AMP were blocked by caffeine. But did that show that caffeine worked through blockade of adenosine receptors? Not at all, says he. Caffeine exerts many effects—on the enzyme phosphodiesterase, say, and even on DNA, the hereditary material. Conceivably, says Snyder, pushing the point to make a point, its effect on DNA somehow explained its pharmacological potency. No one knew for sure.

Work by him and his collaborators, he says, "established rigorously what was happening. I'd had a strong feeling that caffeine worked through adenosine receptors. But we pinned it down."

Snyder admits glycine was a mistake, pure and simple, but

claims his crediting of others' work on enkephalin convertase was well within the bounds of accepted citation practice.

As to the more general charges lodged against him—that he's overly ambitious, insufficiently careful, unmindful of others' toes, and maybe even scientifically unscrupulous—Snyder defends himself through the artful use, as a kind of rhetorical stand-in, of his friend and former colleague, Pedro Cuatrecasas. Snyder knows little of what people say about him, he begins, but he has heard how they talk of Cuatrecasas, whom he lauds as brilliant, accomplished, and creative. Much of what he's heard, he says, bears the unmistakable ring of jealousy.

How, asks Snyder, warming to his argument, can the mediocre scientist, some nameless practitioner of the second rate, reconcile the success of a Cuatrecasas? Well, he can label himself a failure. "Ah, but that will hurt his ego."

It's more natural to say, "I'm serious. I'm careful. But here's this other scientist: He makes up data. He's wicked, evil. He'll rot in hell."

Snyder goes on in like vein: Say you've got two scientists, Aay and Bee, in the same field. Aay discovers a cure for cancer. Bee says, "That son of a bitch. I've been meaning to do that. But Bee *didn't* do it. Aay did."

Yes, any discovery owes much to prior discoveries. After all, "whatever gets discovered doesn't come from Mars." But it's the better scientist, the Cuatrecasas, who, like the chess grandmaster, thinks one step ahead, who actually takes the step—does the experiment, writes the paper, and has it published. The second-rater says, "We were at the same meeting [and heard the same key finding revealed]. That's where he discovered it." Maybe so. But one scientist acted on it, while the other did not.

And yet if simple envy is behind the enmity toward him, Snyder is asked, why don't you hear similar stuff about, say, Julius Axelrod? As a matter of fact, he replies, you do—or at least, you *did*, before Axelrod won the Nobel Prize.

Axelrod is today among "the annointed of God," like any

other Nobelist. But he, Snyder, remembers how back at NIH similar criticisms could be heard about him—that "he really stole COMT. That he'd hear things at a meeting, rush home, do a quick experiment, and rush into print. That he was a slob in the laboratory. . . ."

When he worked for Axelrod, says Snyder, he spent much time defending him. "Then, after he won the Nobel Prize, people shut up." Before then, Axelrod had little to show for his career, not even an office—little, that is, except an astonishing string of scientific breakthroughs. "You see, that was enough to provoke people to say bad things about him."

And with that invocation of the name of Julius Axelrod, Sol Synder rests his case.

•

In 1972, the race was on for the opiate receptor. In 1974, it was for the endogenous ligand.

Endogenous means within the body. A *ligand* is something that links or attaches to something else. The endogenous ligand sought so feverishly during 1974 and 1975 was a substance within the body that linked to something. Linked to what? Linked to the opiate receptor.

Ever since the discovery of the opiate receptor, Snyder and others had been struck by one endlessly seductive implication of its very existence: "We can assume," a *Newsweek* writer had quoted Snyder as saying, "that nature did not put opiate receptors in the brain solely to interact with narcotics." Later, Snyder couldn't recall saying it. But the idea tugged at his imagination. There had to be something, something naturally in the body, some neurotransmitter, perhaps, but in any case, *something*, that worked on these receptors when heroin or morphine didn't; or else, why have receptors in the first place?

That something was the endogenous ligand. And in May 1974, at a meeting of the Neuroscience Research Program held in a stately old mansion in Brookline, Massachusetts, a research group headed by Hans Kosterlitz of Aberdeen, Scotland, announced that they were hot on its trail.

Opiates are known to inhibit the contractions of certain smooth muscles, including the vas deferens, the spermatic duct that carries semen to the tip of the penis. So, using the electrically induced contractions of this muscle as a measure of opiate action, Kosterlitz and his lieutenant, John Hughes, bathed the muscle in partially purified brain extracts, which, sure enough, reduced the contractions. When naloxone was added, the contractions returned, reinforcing the supposition that some opiate, found in the brain extract, was responsible.

At the Brookline meeting, Hughes reviewed what they'd already learned about this natural opiate, this substance that magazine articles would later dub "the brain's own morphine": It did not dissolve in organic solvents such as acetone, but did in methanol and in water. It had an ultraviolet absorption peak at 270 nanometers. Its molecular weight was somewhere between three hundred and seven hundred. And so on.

This was a bombshell indeed. As Sol Snyder recounts it, an earlier discussion with Kosterlitz had piqued his interest in the endogenous ligand question. He and Pert has done some preliminary work, without success. After talking to Kosterlitz, he'd assigned to the problem Gavril Pasternak, who'd conducted some exploratory experiments even before the Brookline meeting. Afterwards, "Gavril moved into high gear."

According to Candace Pert, that is a wild understatement. Following the Brookline meeting, she says, Snyder became obsessed with chasing down the endogenous ligand; even inserted into the record of the conference, of which he was coeditor, lengthy sections concerning Pasternak's early work that, to hear Pert tell it, he hoped might later help establish his scientific claim.

"The moment we got back to the lab, Sol was saying, 'OK, now we go for it.' It was 'Candace, do this. You do that, Gavril. You take this approach. . . .' "

Pert protested, "But what is there to do, Sol? Hughes has got it. It's taken. It's his." But Snyder, by Pert's reckoning, was determined to beat the Scottish researchers, despite their substantial lead. Robert Goodman confirms that an air of fevered competition bubbled through the lab at the time.

It was all to no avail. On December 18, 1975, Kosterlitz and Hughes (along with the chemist Howard Morris) came out with their landmark paper in the distinguished British journal, *Nature*, detailing the five-amino-acid molecular structure of the endogenous ligand they called enkephalin.

"I didn't feel defeated or in any way saddened," Snyder would later record. "Both Hughes and Kosterlitz were and continue to be good friends." Michael Kuhar, now himself a professor in Snyder's Department of Neuroscience at Johns Hopkins, agrees that while "Sol tried to fit it in, and certainly was very interested in it, I don't think he felt beaten on enkephalin. It's not what he was primarily after." Goodman, too, thinks Snyder lacked any great emotional investment in it.

Candace Pert, though, saw it all quite differently. With the race for enkephalin, "I feel he lost his morals. He wasn't the same afterwards." To her, it was a preview of all that was to come.

11. The Lasker Flap

"*THE LAST YEAR* with Sol," says Candace Pert, "was like the adolescent thing, where you grow apart as you start your own house." It was more than four years since she'd joined Sol Snyder's lab. She had her Ph.D. It was time to leave the nest. "Looking back," she says today, "I don't know why I wasn't terrified."

She was twenty-eight, with an international name as co-discoverer of the opiate receptor. But what would she do next?

Whatever it was, she still had Snyder in her corner. "He threw things my way through the old boy network," she says, got her interviews, made her availability known to the right people. For Snyder, it was a personal challenge to help land his students the best jobs—sometimes resorting, Pert adds cryptically, to "fabulous, devious tactics" to do so.

She sent off nine or ten applications, went on interviews, gave guest lectures, all the while keeping her pregnancy with her second child, Vanessa, hidden. The University of Florida at Gainesville, the University of Chicago, and the National Institute of Mental Health all made her offers. She chose NIMH, the lure of its Clinical Center and a parallel offer for husband Agu being the deciding factors. She was named a staff fellow of the Section on Biochemistry and Pharmacology of the Biological Psychiatry Branch.

Arriving in Bethesda in September 1975, she was consigned to a small library area while waiting for her new lab in Building 10 to be equipped. She was "starting from scratch, building my empire," she says in a cocky mood years later. But having left the security of Snyder's lab, she admits, "my insides were clenched. I thought, What am I going to do?"

Until now, she'd always worked in someone else's lab, on someone else's project. Now she was on her own. She was expected to work on the brain, but otherwise had a free rein —so free it was scary. "You did great work here," Snyder had advised her before she left. "Now show you can work on something else." *Steer clear of my turf*, she heard.

There was someone else from whom she sought advice, and even before her first day on the job, she wandered down to his office on the second floor to get it. Five years after winning the Nobel Prize, Julius Axelrod was by now a legendary figure around NIH. The two had what she recalls as a relaxed and pleasant talk. "What are you going to work on?" he asked her. She wasn't sure. "Work on what you know," he advised.

What she knew better than anything was opiates. They were, to her, "like a recipe you've done and you feel good about doing," something familiar. "It was absurd not to work in that," she says. "Besides, I had a million questions."

Pert had never done a proper postdoc, which is normally completed at an institution other than that awarding the Ph.D. She was already a name, so that after an extra year beyond her doctorate in Snyder's lab, she'd moved right into a much coveted NIH slot. There, with receptors and enkephalin the leading edge in the neurosciences and she an acknowledged trailblazer, she soon began attracting students of her own. Among the first was Terry Moody.

Moody had attended the University of California at Berkeley during the tear gas-scented 1960s. For grad school, he'd picked sober-minded California Institute of Technology. For a postdoc, he'd looked to the Baltimore-Washington area, a beehive of neurosciences research. Among the labs to which he applied were Axelrod's, Snyder's, and Pert's.

Axelrod and Snyder couldn't take him, but Pert, just starting up, could. The two of them met for dinner, at a scientific conference they were both attending, to work out the details. He remembers her telling him about NIH's tennis courts—of no minor import to a tennis buff from California.

It was 1977. Candace Pert had her first postdoc.

His first day, he sat down with her to discuss potential projects. After mentioning several, Pert added, "Oh yeah, and there's this bombesin."

"What's that?" he asked.

Bombesin was one small part of the peptide revolution launched by enkephalin. Enkephalin has the molecular structure of a peptide, a short chain of amino acids, of which there are twenty different kinds, strung out like beads on a string. After enkephalin, other peptides began being found in the brain—secretin, substance P, somatostatin, neurotensin, and maybe two dozen more—abruptly complicating most long-held notions of brain neurotransmitter function. Once, there'd been the familiar noradrenaline, serotonin, dopamine. Now, appearing in concentrations too low to be detected before, was a whole soup of transmitters, each with unique properties, each associated with its own neural pathways, adding up to a new chemical language of the brain. And bombesin was part of it.

Bombesin, Moody learned, was a peptide chain of fourteen amino acids that, among other things, lowered body temperature and served as some kind of satiety agent. Later, it was implicated in a particularly deadly form of lung cancer known as oat cell carcinoma. In all, an odd mix of properties. "There's got to be a receptor for this," figured Moody. Anything that potent had to have one.

Pert thought so, too, and to formally demonstrate the bombesin receptor became Moody's project. It took him four months. The trick was finding a way to bind a radioactive isotope to the peptide while retaining its biological activity. He tried four methods, got nowhere, but succeeded on the fifth, using radioactive iodine. "Bombesin: Specific Binding to Rat Brain Membranes," appearing in the *Proceedings of the National Academy of Sciences* in 1978, was the result.

(Later, Candace Pert's father, Robert Beebe, was himself diagnosed with oat cell carcinoma and became a patient at the Clinical Center. While he was treated conventionally, with radiation and chemotherapy, his daughter pursued a long-shot cure, trying to find a toxic agent that might bind to the bombesin receptor and so kill the cancerous cell.

"Hang on, Dad," she'd say when she visited him in the hospital. "A few more days and we'll have the cure."

"Better hurry up," he'd say. In March 1980, he died.)

Moody thrived at NIH, with its tiny, crowded labs and fiercely competitive atmosphere. He stayed two and a half years, before leaving for George Washington University Medical Center and a career built up from bombesin.

His relationship with Pert? "A scientist's ego being what it is," he notes delicately, it was unlikely the two would never clash. Still, they worked well together, his cool methodicalness balanced by her impulsive fire. She was "the most liberal scientist I've ever met," he says. She was always taking experimental long shots, aggressively moving onto the next step. "That," he learned, "is how you discover things, by taking chances."

Pert fancied herself, in her words, "a New Wave scientist, not someone who plays by the old rules like the boys," but rather by her own noncompetitive ones—unlike Sol Snyder, for example, whom she judged "shamelessly competitive." Back at Hopkins she remembers him urging her, "Better hurry, or Eric Simon's going to catch us," referring to a competitor in the opiate receptor race.

The Chapel Hill conference in 1973, with Kosterlitz and the others glorying in her achievement, had supplied her a noncompetitive model she wanted to apply to her own lab: no one pitted against another, as she'd been against Gavril

Pasternak, a free, open intellectual environment. "You can't be secretive," she says. "You have to surround yourself with the smartest people your ego can stand, then concentrate on the work, not on who'll get credit for it."

Yet her vision was streaked with ambivalence. What kept her working, she admitted once, was fear that someone would beat her to a discovery. "Without competition," she wondered out loud, "maybe all you get is creative fluff."

As lab chief, she tried to keep her assistants toiling away in the lab trying things, "not 'watching television,' as Sol would say." No big, elaborate experiments, the nuances and refinements all worked out beforehand. "It's more," as one colleague says of her style, "whatever works"—maximum payoff sought for minimum effort, the experimental protocol, such as it was, perhaps scribbled onto a scrap of paper. It was Julius Axelrod's "skim the cream" tradition all over again, says Agu Pert, who still sometimes works with his ex-wife: Why spend days grinding out reams of data, when with some perfect little coup of an experiment, you could learn just as much, then let others come in to mop up?

"One experiment is worth a week in the library," she remembered Snyder always telling her. So don't think about it too much. *Just get hysterical and do it.*

Pert sees herself as a good bench scientist. Most who have worked with her do not. "She's as sloppy and wild in her lab technique as she is in the way she thinks up ideas," says one otherwise-admirer. Her strength lies in drumming up excitement, cheerleading, and directing, along fanciful paths, those of more workmanlike temperament. "She doesn't think like a scientist," he continues. "She's more the artist, squinting at data, making shapes with her hands when she talks about receptors. To her, a receptor is not a set of numbers, but a living, breathing thing you can sidle up to.

"She just keeps spieling out ideas. She's not into refining them. That's the work of more mundane scientists. Whatever's on her mind comes out. Ninety percent is ridiculous, but the other ten percent is fantastic stuff."

Pert had learned from Snyder to look at data in the most

favorable light. No experiment, after all, ever turns out as planned. The want of hoped-for results can mean a technical glitch as well as an invalid idea. So best to take a second look at your superficially discouraging lab results. "You have to dream from the data," says Pert. Yes, she knows, sometimes her ideas are crazy. "But so what? I don't go into print with them."

Burnout is an occupational hazard among those who have worked with Candace Pert. She is intense, intimidating, and draining in long stretches. "I enjoy working with her," says Agu Pert. "But some people don't, especially men. She can be very authoritarian. 'Do this. Do it my way. Do it now.' Not everyone can handle it."

One who can, and has, is Miles Herkenham, a thin, dark-featured Californian who came to NIH as a staff fellow in 1977. Herkenham had learned of Pert's opiate receptor discovery while still a graduate student at Northeastern, then later about her autoradiography work with Michael Kuhar. She and Kuhar had worked out a way to inject radioactive opiate into the brain of an animal and then, with the animal dead, take a thin slice of its brain and expose a piece of photographic film to it. The resulting image, or autoradiograph, recorded the distribution of the drug in the brain at the time the animal died. Which, with the drug bound to receptors, would thus correspond to an opiate receptor map.

Herkenham, who'd for some time kept one of Pert's autoradiography papers taped to his file cabinet, noticed seeming parallels between her autoradiographs and the neural pathways that, as a classical neuroanatomist, he had traced in the brain. Finding voids in his tracings, he was sure he saw matching areas in hers. I bet opiate receptors fit those holes, he thought.

He wanted to talk to Pert about his idea, yet hesitated. She's a big cheese, he thought, and I'm a little nobody. He was but three years her junior, and held a position only one level below hers. But she was a name in an exploding new field, while he was stuck in a dying one.

Finally, mustering his resolve, he invited her to a talk he

was giving. She came, immediately saw what he was driving at, and has worked with him since.

His first impression of Pert was "how young she was, given how famous she was. My next reaction was how wild and crazy she was. Then, how imaginative, how much a genius she is."

Finally, reality set in. "It hit me that I was going to have to do most of the work. She's more given to generating ideas than to washing dishes."

For two years, Herkenham dropped everything he was doing to work with her, at first largely as her technician, trying to develop autoradiographic methods that would yield images better able to correlate receptor sites with neural pathways. His expertise lay in histology, the microscopic study of anatomical structure; hers was in pharmacology. In many days and nights in the lab, they educated each other. "I know a lot of pharmacology now, all without reading the papers," he says. And she knows a lot of neuroanatomy."

For him, their work together offered a rare chance to bridge classic neuroanatomy with the emerging neuropharmacology of which Pert represented the New Wave. Abruptly, he found himself breathing the heady, rarefield air of a hot new research area. He'd been mired in what had become a stagnant field. "Now, suddenly, I was in the front line."

•

It sometimes seemed to Herkenham that Pert showed an exaggerated range of emotional response, often reacting the same to seemingly trivial things as genuinely important ones. "She was always outraged by this or that," he says. So one day in 1978 when she again seemed angry and upset, this time over some award or another he'd never heard of, he took little note.

Terry Moody remembers the day, too. He was working in the lab, on an experiment, when Pert came in looking upset. Only later did he learn what had happened.

Pert had gotten a call from Sol Snyder. "Guess what?" he

said. "I won an award." It was the Albert Lasker Award for Basic Biomedical Research. In essence, it was the American Nobel Prize. Twenty-eight times before its winners had also won Nobels. Snyder was calling to invite her to the awards ceremony. All sorts of dignitaries would be there, including Senator Edward Kennedy.

"Great, Sol, what's it for?" she remembers asking.

It was for the discovery of the opiate receptor and enkephalin, Snyder replied. Hans Kosterlitz and John Hughes, the discoverers of enkephalin, were sharing it with him.

Kosterlitz and Hughes. Snyder and . . . Pert. Except, this neat parallelism wasn't reflected in the award. What about her? "I went wild," Pert remembers. She had discovered the opiate receptor, yet he was to be honored for it. She was incensed.

"I was surprised at her reaction," says Snyder. "I tried to calm her down."

" 'You know and I know who did the crucial work,' " Pert remembers Snyder assuring her.

But nothing he said consoled her. She could understand if Snyder and Kosterlitz alone had been honored. "Then I would have gone to that luncheon and beamed with pride for Sol. I'd be right up there shaking Sol's hand." It is the rule in science, after all, not the exception, that senior investigators receive honors and acclaim based on work physically done by junior colleagues.

But if that were so, if that was the principle invoked, why then was Hughes getting a piece of the Lasker? Didn't he bear the same relationship to Kosterlitz as she did to Snyder? And if he got it, why didn't she?

"Candace, you know I had nothing to do with it," Snyder protested to her. Publicly, he said that "it would have been appropriate if Pert had shared the award," though he understood how the Lasker jury might arrive at a contrary judgment. The Kosterlitz-is-to-Hughes Snyder-is-to-Pert argument won him over as reasonable. "Let me call the people I know on the Lasker committee," he remembers telling her.

When he did, he was told the committee had specifically considered several other names, Avram Goldstein and Candace Pert among them. However, it had duly weighed its decision, and that was that. Snyder reported the conversation to Pert. "She acted like I was personally doing this to her," he says, throwing up his hands in frustration. "It was almost like I was God or something."

In the weeks between that first phone call and the formal luncheon in November at which the award was to be presented, Pert was on the phone with Snyder constantly. She says she tried to bargain with him, asking him to turn down the award in protest; or else to publicly give half the award money to Bryn Mawr, her alma mater. Snyder's response? Pert says he "stonewalled, denied. He went around and around. I'd come back to it, and he'd say, 'Candace, let me explain it another way.' "

Her anger and hurt persisted. She had received a formal invitation to the award ceremony. She didn't respond. "Please —We have not as yet received your reply card for the Lasker Award dinner, Tuesday, November 21, 1978," a postcard reminded her. A year later it was still taped over her desk. She never went.

She knew that if she did go she'd be there smiling. But she just couldn't do it. And so, a week before the award ceremony, she wrote Mary Lasker, husband of the late Albert Lasker and a powerful force in American medical research. "I was angry and upset to be excluded from this year's award," she wrote. "As Dr. Snyder's graduate student, I played a key role in initiating the research and following it up."

That excerpt from her letter surfaced early the following year in the pages of *Science*, under the headline, "Lasker Award Stirs Controversy." A little later, a reporter for *Science News*, Joan Arehart-Treichel, dug into the story, concluding that sexism had been a major factor in Pert's exclusion, but that the sin, such as it was, was more of omission than of commission. "But then," she added, "doesn't sexism usually work that way?"

The *Science News* article caused much embarrassment, excited much comment. The case became a *cause célèbre* among feminists, Topic A among scientists for months. Feelings were sharply polarized, with some seeing blatant sexism at work, while others saw the episode as an ugly blemish that had no business being bared. Pert earned instant notoriety, being introduced at one lecture, for example, as "the Scarlet Lady of Neuroscience."

She had done the unthinkable, airing openly what she regarded as an injustice, brusquely pulling aside the veil that obscured science from the public. She had made plain for all to see that behind its veneer of cool reason, science could be just as messy, just as ugly as any other realm of human affairs, and could whip up passions quite as hot.

•

"One of the ongoing mysteries of Nobeldom swirls about Solomon Snyder. Why hasn't he won?" asked Hearst newspapers science writer Joann Rodgers in 1983, not long before she was named deputy head of public relations at the Johns Hopkins Medical Institutions. The reason, she said, was Candace Pert.

"Snyder has picked up other international prizes for his brain research and dazzles colleagues here and abroad with a dizzying array of experiments that always seem to strike gold. . . . If there is a reason why Snyder has yet to walk across the stage of the Stockholm Concert Hall, it probably has to do with a particularly quarrelsome interlude precipitated by a former graduate student and colleague, Candace Pert."

The "quarrelsome interlude," of course, was the Lasker flap.

Stockholm-watchers have long noted that one sure way to remove yourself from Nobel contention is to become touched by so much as a breath of scandal. As William K. Stuckey, writing in *Omni*, has put it, "The austere Swedish message is to shut up, do your science, and keep your nose clean." The

Lasker mess, Rodgers was suggesting, had left Snyder tainted in the eyes of the Nobel committee.

Rodgers was not alone in her assessment. Nobel Prizes are normally announced on set days of the year. But the year after the Snyder-Kosterlitz-Hughes Lasker Award, they were delayed by a mysterious debate within the Nobel assembly. Ultimately, the prize honored the developers of the CAT scan. "It was a kind of engineering prize," sneers Edith Hendley, for a technical development which, however clinically important, did little to advance fundamental knowledge. Enough people on the Nobel committee were put off by Pert's outcry, she speculates, to scotch the Snyder Nobel at the last minute.

Pert herself admits little doubt. "Sol," she declares unequivocally, "got stopped by me from getting it."

Snyder supporters were incensed by the Lasker ugliness, invariably describing Pert as ungrateful for all Snyder had done for her. "I've never been so angry at Candace in my life," says Pasternak—which, given their past relationship, is saying quite a lot. He feels she was "totally, unequivocally, absolutely, all wrong. I contributed as much to Sol's Lasker, and I don't feel I deserved a piece of it; she deserved it even less."

The opiate receptor was "world class stuff" and, to his way of thinking, Snyder erred in letting Pert, a second-year grad student, present it to the world. There she was, fielding questions from the press, seeing herself in *Newsweek*. "Candace got the reputation as an internationally known scientist," he snaps his fingers, "like that. And that's hard to live up to. It can give you delusions of grandeur. I think Candace succumbed. She believed what people said about her."

Michael Kuhar came to a similar conclusion, but in a different way. Snyder, he says, has a knack for leading you toward a scientific insight. You'll finally see what he's driving at. He'll exclaim, "Why, that's it! You got it!" And you'll come away sure you'd thought of it yourself. In just such a way, he suspects, Pert came away from the opiate receptor discovery convinced she'd done it all herself. "My own reaction," says Kuhar, "was that it was amazing that a grad student should think she ought to win the Lasker Award. It was, 'Holy cow, this is a little much!' "

Pert, in any case, felt wounded, hurt, misunderstood. She didn't come into the lab much for a while, remembers Terry Moody. Around this same time, she slashed her hand on a glass door at home, and when she did show up she was in much pain. Moody remembers thinking that the mental and physical pain fed off each other.

Before the Lasker affair broke, Pert had been on the phone with Snyder all the time. "She wouldn't hold anything back from him," Moody says. "She viewed him literally as a father figure." Then, as the Lasker controversy erupted, their relationship soured. Around the lab, Pert rarely spoke of it, hid her hurt as best she could. Still, to Moody it was evident that she was "distraught about it, obsessed."

A woman friend of both Pert and Snyder, Merrily Poth, tells how, during the thick of it, she and Pert sat in a health club sauna, Pert painting the Lasker controversy as an ethical issue, arguing that just as gentiles in anti-Semitic Nazi Germany faced inescapable moral decisions, so now was unambiguous choice a moral necessity. "You have to choose," she insisted. "You've got to pick me or Sol."

Before she'd written Mary Lasker, says Pert, "I gave Sol every opportunity. I wanted him to at least say, 'Sorry, I'm a regular prick. But you're a smart chick. Let's make up and be friends.' " But there was no concession. "Sol was never *mensch* enough to make a single gesture." And that, she says, forced her to write Mary Lasker. "It was the hardest thing I ever did."

And one of the most personally meaningful. "The 'right' thing for me to do would have been to curl up and get cancer, like Rozzie Franklin," she says. Rosalind Franklin was the British X-ray crystallographer who helped unravel the structure of DNA, and died in 1958, at the age of thirty-seven, without sharing in the acclaim enjoyed by James Watson and Francis Crick. She has been a figure of controversy since.

In Pert's view, to do nothing would have amounted to scientific death. The Lasker Award, as it stood, "made me a total nonentity. I knew I wasn't."

•

While most scientists were appalled by how Pert aired her grievances, no similar consensus applies to the actual merits of her case. Some scientists felt she deserved a share of the award. Some assigned her abundant credit for most of the work and many of the insights leading to the discovery—but then concluded that she was just a graduate student, for God's sake, and graduate students don't get Lasker awards. Finally, some felt that her case, even on narrow scientific merits, was groundless—that Snyder was well on his way to the opiate receptor no matter which grad student happened to be floating through his lab at the time, and that Pert was little more than an instrument of his genius, hands to execute the ideas bubbling forth from his fertile imagination.

While Edith Hendley agrees with most others that it was unwise for her to go public with her case, she nonetheless accepts Pert's claim on the Lasker. She regards as convincing, first of all, Pert's claim of parity with John Hughes—that if he, as junior to Kosterlitz, got a piece of the Lasker, she, as junior to Snyder, deserved a piece of it, too. Pert's thinking was "perfectly reasonable," she says, "and it must have struck Sol as perfectly reasonable, too, because he [later] tried to get her included."

But wasn't Hughes, in fact, no graduate student at all but scientifically on his own? And didn't one of the enkephalin papers, a fourteen-pager in *Brain Research*, appear under his name alone, without Kosterlitz? To Hendley, it makes no difference: Hughes was clearly junior, just as Pert was. (Kosterlitz was well aware that, as an opiates researcher of substantial reputation, he might overshadow Hughes, and for that reason, he told Eugene Garfield of the Institute for Scientific Information (ISI) in 1979, relinquished authorship of the *Brain Research* paper. In his column in the ISI periodical *Current Contents*, Garfield sympathized with Kosterlitz's wish to help a colleague, but decried the practice in any case.)

Hendley says that long before Pert arrived in the lab, Snyder was talking about the opiate receptor; "just to give it a name" was, in her view, critical. Moreover, she feels sure the discovery would never have happened outside his lab. Snyder, once he'd seen Goldstein's paper, "picked up on it and went for it like a bat out of hell," she says, guiding Pert's thinking all the way. "I don't think Pert would have done it without him."

But, she stresses—and to her this is the decisive point—Pert *did* do it.

Several pieces of evidence suggest that doing it was, as a technical problem, no matter of routine recipe-following for which success was taken for granted.

First, Avram Goldstein, though he'd drawn a route map toward the opiate receptor, never succeeded in traveling it himself; the map was but a crude one, with many an obscure turn along the way, and across difficult terrain. From when Goldstein submitted his 1971 paper to when Pert and Snyder reported success in *Science*, almost two years elapsed.

Second, Snyder notes that in an NIH grant proposal he submitted at the time, he deliberately underplayed his interest in the opiate receptor, despite his own fascination with the problem, because he knew the funding authorities would see it as too chancy. Even the brief mention he did make, he reports, was greeted as "a most risky flier."

Third, though included in a pile of potential student research projects that Snyder kept on his desk, the opiate receptor repeatedly drifted to the bottom—because, as Snyder wrote in an account of the discovery, "the project was more of a long shot than the average experimental endeavor." When he pulled it out at last, he gave it to Candace Pert.

Finally, when initial experiments seemed to be leading nowhere, Snyder was discouraged enough to want to drop it altogether, at least for a while. Pert wanted to give it another try. She did, with naloxone, and succeeded.

Pedro Cuatrecasas agrees that the opiate receptor was no trivial problem. But he points out that Pert's work did not emerge from an intellectual vacuum, that she got lots of help from others around the lab—on pH, on what buffers to use, on how much radioactive ligand to try, and so on. He agrees she played a key role; she listened, synthesized from what she heard, pursed her goal relentlessly. Still, he says, "it wasn't as if she were isolated in that lab alone.

"Sol clearly deserves the credit. Candace did important work, but I don't think she's responsible for the real original part of it. It's not just the idea, but the push for it that came from Sol."

•

Six months after the Lasker Award presentation, Eugene Garfield, of the Institute for Scientific Information in Philadelphia, grappled with the issue from quite a different perspective. Garfield is the foremost champion of citation analysis, a quasi-mathematical tool based on the custom, within science, of acknowledging in print the insights, theories, suggestions, and evidence contributed by others.

Every scientific paper credits papers preceding it. Keyed by footnotes scattered through the text, cited journal articles are listed at the end of the paper, where they appear in a form something like this:

5. Van Praag, D., Simon E. J.: Studies on the intracellular distribution and tissue binding of dihydromorphine-7,8-^{3}H in the rat. *Proceedings of the Society for Experimental Biology and Medicine.* 122:6–11, 1966.

6. Goldstein A., Lowney L. I., Pal B. K.: Stereospecific and nonspecific interactions of the morphine congener levorphanol in subcellular fractions of mouse brain. *Proceedings of the National Academy of Science.* 68:1742–1747, 1971.

7. Pert C. B., Snyder S. H.: Opiate receptor: Demonstration in nervous tissue. *Science.* 179:1011–1014, 1973.

And so on. There might appear half a dozen such citations, or twenty, or occasionally, as in a review article surveying a whole field, a hundred or more. This References and Notes section, as it's apt to be called, is the bane of typists—most prone to error, most tedious to prepare. And yet it's crucial, for it forges the intellectual link between this latest finding and all that's come before, aptly expressing Sir Isaac Newton's famous statement of scientific humility and interdependence: "If I have seen further, it is by standing on the shoulders of giants."

Some giants, however, are taller than others, their shoulders affording a more commanding view of the scientific terrain.

Most scientific papers, let it be said, are never heard from again; they contribute nothing about which anyone cares, change no prevailing ideas, provoke no new thinking. By one count, half of all papers are not cited even once in the first year after their appearance. A few papers, on the other hand, are cited again and again, and so are seen as making the greatest contribution to their fields.

Starting from this premise, Garfield looked into the 1978 Lasker Awards, gathering data on the papers of all those involved in the initial quest for the opiate receptor, fashioning maps of scientific influence. His conclusions lent credence, first of all, to a point made by Thomas H. Maren of the University of Florida at Gainesville in a letter to *Science*. Why, Maren wondered, had Avram Goldstein been excluded from a share in the Lasker? And what about Lars Terenius, of Uppsala University in Sweden, and Eric Simon of New York University, whose near-simultaneous demonstrations of the opiate receptor had somehow been lost in the shuffle? "All of this work is inextricably linked."

That's just what Garfield's analysis showed. Goldstein's paper was "of prime importance to research on opiate receptors," Garfield wrote. It came early. It was unique. It paved the way. But the other key figures in receptor research could likewise be defended. "Each of these scientists has a strong claim on the discovery."

Indeed, a National Institute of Drug Abuse (NIDA) Award a year prior to the Laskers went to Simon, Goldstein, and Terenius in addition to the three future Lasker winners. While Pert was again excluded, NIDA's William Pollin two years later wrote *Science* with what amounted to an apology: "In retrospect, we feel it was a significant omission on our part that Dr. Candace Pert was not included. Her graduate student role was the issue at the time; subsequent increased awareness of her major contribution has led us to this revised conclusion."

A contribution major enough to merit the Lasker? Garfield found that based on her work following the discovery of the opiate receptor, Pert continued to be "the only coauthor of any of the senior investigators to appear so frequently" in the citation maps. These later papers showed "that she is still a force within the specialty without the help of her mentor." Moreover, the seventeen papers she and Snyder coauthored from 1973 to 1976 received an average of eighty-seven citations per article, while those Snyder published with other collaborators got just thirty-eight per article. The evidence, Garfield concluded, couldn't prove Pert's contribution to the opiate receptor work was crucial, but at least hinted that she was capable of it.

"Both the cluster data and citation counts," he claimed, "provide strong evidence that Candace Pert deserves formal recognition for her contributions."

•

For a few years, Donald Brown lost track of Snyder, his old guitar teacher from NIH days. But then, during the 1960s, they both wound up in Baltimore, Brown at the Carnegie Institution of Washington's department of embryology, Snyder across town at Hopkins. Becoming first reacquainted, then friends, they today live in the same north Baltimore neighborhood, Mount Washington. Their wives and kids see each other. They follow one another's careers—though Brown says he's familiar with Snyder's work "no more than if he were a physicist." What he likes best about the younger man is that "with Sol, what you see is what you get. In any role I've ever seen him in, he's been the same guy."

The Lasker episode greatly upset Snyder, he reports. "I didn't award myself the prize," Snyder told him. Brown, who first heard Pert's displeasure voiced in the pages of *Science*, thought her reaction absurd, and remains sympathetic to Snyder. Had Pert worked down a floor, in some other lab, he asks, could she have done it? He doesn't think so.

A senior scientist who himself directs a large laboratory, Brown, fifty-three, sees graduate students as virtually the creations of their lab chief. Students come into a lab like his, or Snyder's, where everything has been set up for them. "They start out with a refrigerator stocked with the latest reagents and lots of grant money. It's a lot different from going to East Nebraska Normal. For young people coming here, it's the greatest years of their lives. They'll never be more productive. The only thing they don't do is select the project. That's the last critical step—selecting what to do."

And for some, the most difficult one. "That is what growing up in science is all about," says Brown. And that's what Pert, while still a student in Snyder's lab, had not yet done.

After leaving Snyder, Pert went straight into a lab of her own. That she could, declares Brown, is a testament to American science's willingness to reward talent regardless of age or experience. Pert should be grateful. "If she were living in Japan, it would be thirty years before she had an independent position. Here, she has her own students, her own lab, her own problems. Now we'll see if she's as talented as she apparently thinks she is."

•

How, then, to weigh the relative contributions of two researchers to a single finding? How to decide whose is prizeworthy and whose is not?

It is true that the opiate receptor as a problem had been the beneficiary of Snyder's thought since at least the day, in the summer of 1971, when he heard Avram Goldstein give his talk; and that Pert first tackled it at his urging and under his guidance. It may also be true that she was uniquely driven to attack it because of her personal experience with opiate drugs.

Success owed much to the rapid filtering technique developed by Pedro Cuatrecasas, and Pert's stint in his lab left her well-versed in its use. But so was Snyder, who worked down the hall from Cuatrecasas and collaborated with him on a related project.

As her critics point out, Pert may not have made the discovery had she worked in some other lab. On the other hand, she'd picked Snyder as much as he picked her, seeking him out in large part because of his molecular approach to the workings of the mind. Furthermore, she had numerous opportunities to swear off the opiate receptor as a scientific challenge, and didn't take them.

It is true that, as a laboratory problem, the opiate receptor was not routine, not trivial, and that Pert brought to it considerable intelligence and tenacity. But it is likewise true that once the Goldstein strategy for an attack on the problem was in place, success was probably a matter of time.

Where, then, does the balance lie? What is justice?

•

There were close to forty of them, the men in jackets and ties, the women in dresses, smiling *cheese* for the camera. Snyder sat in the front row, right leg slung easily over the left, head cocked to the side, his students fanning out in rows behind him. In the back, standing next to Agu and just behind Gavril Pasternak, was Candace Pert.

The time: November 4, 1979, the furor surrounding Snyder's Lasker Award less than a year past. The place: Atlanta, at a meeting of the six-thousand-member Society for Neuroscience, of which Snyder had just been named president. The occasion: A group of Snyder's present and former students had met to honor him.

They rented a room in a Peachtree Center restaurant, the Midnight Sun, and coughed up sixteen dollars each for an elegant buffet dinner. The fear was that Snyder might prove unable to come at the last minute, recalls David Bylund, who organized the event, so the honoree knew about it from the start. What he didn't know was that his wife Elaine was being flown down from Baltimore to join him.

At the dinner, Snyder was presented reprints of all his scientific articles, some four hundred of them, handsomely bound into four thick volumes. "It was a stack of books this high," says Gavril Pasternak, lifting his hands a foot off the desk.

In a preface to the bound volumes, Edith Hendley wrote how, while it was customary to honor a scientist after twenty-five or fifty years of research, Snyder forced a contraction of the usual time scale; he had only just turned forty. "It boggles the mind to reflect on the best that is yet to come," noted Hendley. But in the meantime, "over forty of us, spread over five continents, have responded with unbounded enthusiasm at the opportunity to honor our esteemed mentor, and to present him with these volumes of his collected works as a memento of shared discoveries, and as a token of our undying affection, admiration, and gratitude."

Each student had submitted a photo of him- or herself for mounting in a collage, which was also handsomely framed and presented to Snyder. Today, it hangs prominently in his office, where he need only lift his eyes to see it. The Atlanta tribute, he says, "was more meaningful to me than winning the Lasker."

Pert sat right there at the head table with him; some grumbling could be heard about that. There'd been some question about whether to invite her at all. "The consensus," says David Bylund, "was that she'd made a fool of herself." But also that she couldn't very well not be invited.

All through the evening, Edith Hendley remembers, Pert tried to patch things over. Bylund delivers a more acid assessment, comparing Pert to the Hungarian diction teacher in *My Fair Lady* who "oozes charm from every pore." While Snyder apparently tolerated her, his wife Elaine did not. At one point, Pert went up to her and said hello, recalls Hendley, who was standing nearby. Replied Mrs. Snyder: "Well, I'm not glad you're here."

"Elaine, cool it," someone overheard Snyder telling her later.

But the tensions between the two women did little to cloud

an otherwise pleasant evening—which, in part thanks to Key Dismukes, a postdoc in Snyder's lab between 1971 and 1973, had its moments of hilarity. Dismukes got up to announce that certain scriptural documents had recently been unearthed, "a newly discovered batch of Dead Sea Scrolls . . . compiled by a group of scholars [whose] spiritual leader apparently was a man named Solomon.

" 'On the morning of the fifth day, God created the synapse,' " he read from one of these "documents." " 'And the synapse was with form and void, and darkness was upon the synaptic cleft. And God said, let there be molecules which are released into the synaptic void, and He named the molecules *transmitters*. And God saw the release of transmitters, that it was good, but it was not enough, so He created the receptor.' "

Dismukes told how God had revealed to the prophet Avram how to identify the true opiate receptor, but how Avram " 'became impatient with the Lord and used ligands of insufficient specific activity and so did he fail. . . . And so it happened that Avram did not receive the Lasker Award.' "

Back then, Dismukes went on, there was " 'a virgin, who was called Candace. And the angel of the Lord came unto her and said, thou shalt conceive and deliver a receptor for opiates. And the virgin Candace became great with child.' "

Some theologians later claimed, Dismukes continued, still in character, that Candace's was an immaculate conception, prompting much controversy. But scholarly evidence had now revealed, he said, "a trail of radioactivity from the lab bench of this person Candace, down the hall, and into the office of the man named Solomon. . . ."

Finally, Dismukes told how one night the Lord's messenger had set off with a revelation for Solomon about how to find the brain's own opiate, enkephalin. But en route to Baltimore he got lost, stopping instead in Aberdeen, Scotland.

"The rest," Dismukes concluded, "is history."

•

She's "sick of talking about it," Candace Pert once told a magazine about the Lasker controversy. She did not want to be known as "a grumbling lady scientist who had done nothing else since then. I have no feud with Sol, whom I respect very much." Asked on one occasion to discuss the controversy, she shot back, "What controversy?" Another time, she shrugged off the whole affair. What did it matter? she said. "One day we'll all be there in Stockholm."

Publicly, Pert has tried to put the Lasker flap behind her. Privately, it pains her still. One moment she'll dismiss the whole episode, ask to drop it, then in the next, come back to it, plainly preoccupied, still hurt, trying to justify herself, to explain, to understand. "I've forgiven him. I don't have the rancor," she'll say. Pause. "Or do I?"

She does.

The story she tells of betrayal and broken trust invites incredulity. But she insists it's true. "I wouldn't lie. My life is too complicated to lie."

She tells of Kosterlitz's young colleague, John Hughes, at one point visiting her and Agu at their house in Bethesda and idly wondering, while they all sat and talked in the living room, how she would feel if Sol won the Lasker without her.

She tells of another visit by a "close lieutenant" of Snyder, whose name she won't divulge, who said he'd seen Lasker nomination materials in the office of Snyder's secretary, when the nomination was supposed to have been the work of department chairman Thomas August.

No, her omission from the Lasker was not the work of impersonal forces, of institutional disregard of graduate students, or anything like it. Rather, Snyder himself had deprived her of it.

Why should he? What had he to gain? If, somehow, he could influence it, why not have the Lasker go to the four of them?

Ah, says Pert, but the Nobel Prize is limited to three; the Lasker is just a steppingstone. Why, not once in twenty years has it gone to more than three; that's how Mary Lasker wanted it. So don't you see? Setting up a neat threesome for the Nobel meant excluding one of the four. She was the easiest to exclude; after all, she was just a woman. Had it been Gavril Pasternak, or some other man, "they wouldn't have dared."

At one point, she tells you, Julius Axelrod called her into his office. "I want you to help me prepare Nobel Prize nominations," he told her. Inwardly, she rejoiced. He meant to include her! But no, Axelrod wanted to nominate Sol Snyder, along with Hughes and Kosterlitz. And he wanted her help. She refused.

"But Sol loves you," Pert has Axelrod saying. "You do this for Sol and he'll help you later. That's how science recognition works."

Once, she'd regarded "Sol as a god, and Julie as the god of gods." Yet now even Julie was excluding her. That was the final indignity, the cruelest disappointment of all.

•

Thomas August, chairman of the department of pharmacology and experimental therapeutics at Johns Hopkins, concurs with none of Pert's account.

He was new in the job when he submitted that year's Lasker nomination. But Snyder, the boy wonder of the pharmacology department, as well as of pharmacology generally, was the obvious choice. Everyone in the department said so. He'd already received numerous awards, had a tremendous reputation.

Pert? "I knew nothing of her," and didn't consider nominating her. Moreover, now that he is familiar with her work, he would have done things no differently, he says.

And yes, he prepared the nomination, not Snyder. It's common, of course, to nominate friends and close colleagues for important awards, and for the nominee to furnish supporting data, some of it highly technical. Who better to supply the dates and details of a scientific discovery than the discoverer himself? And so, as August says, "I obviously had a piece of paper that described Sol's work." But so did he of others in the department.

The nomination itself came from his office, he insists, not Snyder's. And it was for Snyder alone—not Snyder and his ultimate cowinners. So much for conspiracy theories.

For his part, Sol Snyder coolly denies any part in preparing the Lasker nomination. Perhaps Tom August at some point solicited background material from his secretary. If so, he knows nothing of it. As to Pert's account to the contrary, "I have no idea what she's talking about."

Nevertheless, Candace Pert asserts today that Snyder had a hand in denying her the Lasker, her impression being the product, she declares, of "hours and hours" of conversation with him. Had he truly played no part, "I think he would have just told me. He would have said, 'God, Candace, you've got it all wrong.' But he never did."

The thought wounds her—that he, Snyder, whom she once "adored," for whom she felt such love, could stoop to such deceit. Snyder's supporters hate her? She knows it: "They feel I betrayed him, but it was he who betrayed me."

I had spent the afternoon with Pert out on the back patio of her little cottage of a house on Custer Road in Bethesda. She had started out with all her usual brashness, her theatricality, her rhetorical flourishes. She'd allowed as how she could never forgive Snyder for what she deemed his ethical transgressions and so on—all the old resentment from five years before, still fresh and bilious, gushing forth.

But now, as she lay out on a beach chair under the warm spring sun, the anger dissipated, giving way to a wistfulness, a quiet sadness I'd never seen in her before. Now she was no longer neuropharmacology's *enfant terrible*. Her theatricality disappeared. Her body grew still, more relaxed. She spoke more earnestly, her voice grown softer, sometimes catching with emotion.

She had come to Snyder straight from Bryn Mawr, with its starry-eyed delight in the pleasures of the intellect and the Search for Truth. At first, she'd noticed no flaws in him.

When she did, they seemed to appear all at once. First there was the way he'd moved in on enkephalin after the Brookline meeting. "I was mad at how he tried to steal it from John Hughes," she says. Then, later, he'd been asked to write a *Scientific American* article about opiate receptors and, though she'd "begged" to coauthor it with him, he'd refused.

Just a few years before, in 1974, she had led off her doctoral dissertation with this acknowledgment of her debt to Snyder. He was, she wrote,

> a dedicated teacher with an uncanny talent for defining the critical scientific question, designing the "right" experiment, disregarding the irrelevant or misleading result, and drawing the essential conclusion. For these reasons, it has been an extraordinary privilege to learn about research from him. I am grateful to Sol, not only for all he has taught me, but also for the extreme kindness, consideration, and generosity which he has extended to me over the past four years.

Her esteem for him was abundantly reciprocated. Right up to the opening salvos of the Lasker war, one who knows them both recalls, "he was totally caught up with her, saw her as the greatest thing since sliced bread. 'She's wonderful, very smart. Everybody in the lab is jealous.' " Then, during the opiate receptor fever, she'd spent hours on end with him. "I had as much access to him as I wanted."

It had been so good, and now they were reduced to this. She and Snyder were never lovers, yet their relationship had been, in her words, "incredibly romantic, truly a scientific affair.

"That was the tragedy," she says. "We loved and respected each other very much."

12. The Mentor Chain

LIKE ALL LOVE RELATIONSHIPS, the course of a mentor relationship is rarely smooth and its ending is often painful. . . . There is plenty of room for exploitation, undercutting, envy, smothering, and oppressive control on the part of the mentor, and for greedy demanding, clinging admiration, self-denying gratitude, and arrogant ingratitude on the part of the recipient. It is not always clear who is doing what for whom. After the relationship has been terminated, both parties are susceptible to the most intense feelings of admiration and contempt, appreciation and resentment, grief, rage, bitterness and relief—just as in the wake of any significant love relationship.

—*The Seasons of a Man's Life*, by Daniel J. Levinson, with Charlotte N. Darrow, Edward B. Klein, Maria H. Levinson, Braxton McKee

Daniel Levinson and his colleagues were not writing specifically about Candace Pert and Sol Snyder—but they could have been. Nor did they have in mind the painful split between Julius Axelrod and Steve Brodie—but, again, they could just as well have. Levinson is a Yale psychologist who, with his coworkers, spent ten years tracing patterns of adult development. (Their study, for methodological reasons, was confined to men.) They discovered, among other things, that intense, emotionally charged mentor relationships, like that between Pert and Snyder, play a key role in men's lives.

The word itself goes all the way back to Homer, who has the departing Odysseus entrust his son, Telemachus, to "faithful and wise" Mentor. Only recently, however, has mentoring become something of a pop phenomenon, touted in everything from academic journals to comic strips, its role in the arts, the professions, and business studied and scrutinized. "Everybody Who Makes It Has a Mentor," proclaimed a headline in the *Harvard Business Review.* "A Conceptual Analysis of the Mentor Relationship in the Career Development of Women" was the title of an article in an obscure scholarly journal. "Do You Need a Mentor?" asked *Mademoiselle* of its teenaged readers.

By now, the mentor idea has burrowed deep into popular culture. A black college president, Benjamin E. Mays, dies and the Associated Press obituary inscribes him as the "spiritual mentor" of Martin Luther King, Jr. William Safire, the *New York Times* columnist, devotes part of a column to the word's origins and proper use. In the "Doonesbury" comic strip, a corporate climber informs his wife he's been voted chairman of the board: "I'm one of the big boys now," he says, basking in self-importance. "I've got to call my mentor." (And he calls back to her from the phone, "What's the number over at the nursing home?")

The Daniel Levinson study of adult male development that led to *The Seasons of a Man's Life* (and on which the Gail Sheehy bestseller *Passages*, was also largely based) yielded many revealing insights into the mentor relationship. Levinson and his colleagues found that men go through predictable, age-governed stages of development, just as children do. For example, when a man is between seventeen and twenty-two, give or take a year or two at either end, he passes through an Early Adult Transition marked by a testing of new waters in every area of his life. During his twenties, he first starts building intimate relationships, makes a provisional job choice, begins Entering the Adult World. And so on, all the way into his fifties and sixties, with striking predictability, each life phase presenting unique developmental tasks.

One crucial task of early adulthood, Levinson found, is finding a mentor. A major satisfaction during midlife is being one.

The mentor functions as a mixture of parent and peer. Typically older than his protégé by about half a generation,

> he may act as a teacher to enhance the young man's skills and intellectual development. He may use his influence to facilitate the young man's advancement. He may be a host and guide, welcoming the initiate into a new occupational and social world and acquainting him with its values, customs, resources, and cast of characters. Through his own virtues, achievements, and way of living, the mentor may be an exemplar that the protégé can admire and seek to emulate.

Most of all, the mentor "fosters the young adult's development by believing in him, sharing the youthful Dream, and giving it his blessing."

•

Science is sometimes imagined as a great reservoir of knowledge growing progressively bigger as one largely interchangeable scientist after another tosses his or her contribution into the communal pot. Or, as Lord Florey, a past president of Britain's Royal Society, once had it,

> science is rarely advanced by what is known in current jargon as a "breakthrough," rather does our increasing knowledge depend on the activity of thousands of our colleagues throughout the world who add small points to what will eventually become a splendid picture much in the same way the Pointillistes built up their extremely beautiful canvasses.

Today, however, most observers of science do not accept this view, holding instead that a few scientists contribute out of all proportion to their numbers; that science amounts to two different worlds—one practiced by a large rank-and-file, the other by a tiny elite. A few top scientists, they point out, discover vastly more, and publish vastly more, than most other scientists. Half of all scientific papers, it has been estimated, are the work of just ten or fifteen percent of all scientists. And the work of this prolific elite counts for more, too. Their papers make a bigger splash, being cited much more frequently—twenty, thirty, or forty times more frequently—than average.

In 1962, sociologist Thomas Kuhn wrote an immensely provocative book, *The Structure of Scientific Revolutions*, that almost overnight altered the prevailing view of science's workings and proved so seductive that scholars in economics, political science, and sociology were soon applying his insights to their own fields. In it he declared that science does not, as most people believed it did, progress through the orderly accretion of neutral fact, theory being adjusted this way or that to accommodate new evidence as it develops. What really happens is that some long-prevailing view of nature undergoes, abruptly, a "paradigm shift"—a scientific revolution in many respects resembling a political one. Einstein's relativity theory, for example, changed the kinds of experiments physicists perform, the instruments they use, the questions they ask, even the types of problems considered important. Einstein ushered in a revolution. So did Newton, Lavoisier, Dalton.

But while the old paradigm yet prevails, most research is based firmly on past breakthroughs; it seeks particular kinds of facts to fit particular gaps of knowledge, employing particular kinds of scientific apparatus. This Kuhn termed normal science, which is what occupies most rank-and-file scientists most of the time.

In the strictest sense, the kind of revolutionary science Kuhn had in mind comes along maybe once or twice a century; as sociologists of science Jonathan and Stephen Cole have noted, even Nobel laureates and others in the highest strata of science may, viewed in historical context, come to rank only as "the bricklayers rather than the architects of science." Still, Kuhn's idea may profitably be extended to smaller-scale revolutions: When Brodie, influenced by James Shannon, began measuring drug concentrations in blood plasma, instead of just noting the size of the dose administered, he was firing the first shots of the revolution that was the New Pharmacology. When Axelrod began to sort out the complex sequence of events taking place at the neuronal synapse, he was helping to usher in the new field of neuropharmacology. When Snyder and Pert—however one apportions the credit—demonstrated the opiate receptor and created a powerful new technology for studying it, they were starting a revolution whose rumble still is heard.

Each asked new questions, used new methodologies to ask them, and left a vast range of follow-up questions for others to pursue. Each, in Kuhn's terms, was performing a species of revolutionary science. And each was staking a claim for a place in science's elite.

In 1979, working from a computer file of some 67 million references and 5 million papers published between 1965 and 1978, Eugene Garfield's Institute for Scientific Information prepared a cross-disciplinary listing of the thousand most cited scientists, their fields ranging from astronomy and astrophysics to organic chemistry, molecular biology, pharmacology, and virology. Perhaps half a million men and women throughout the world, it has been reckoned, at least occasionally publish scientific papers. So the ISI one thousand represented illustrious company indeed. Even the membership of the world's most prestigious national academies, like the U.S. National Academy of Sciences, number many more than that.

Among those high on the ISI list were Steve Brodie, Julius Axelrod, and Sol Snyder. Others, from Brodie's lab, were Sidney Udenfriend, Herbert Weissbach, James Gillette, Arvid Carlsson, Elliot Vesell, Alfred Pletscher, and Erminio

Costa. From Axelrod's lab: John Daly, Leslie Iversen, Jacques Glowinski, Irwin Kopin, Richard Wurtman, and Hans Thoenen. And from Snyder's relatively young lab: Michael Kuhar and Diane Russell.

Others on the list included Bernard Witkop, Gordon Tomkins, Christian Anfinsen, Donald Brown, Jack Orloff, Fred Goodwin, Avram Goldstein, and Pedro Cuatrecasas. All, by this one, exacting standard, qualified as among the elite.

There are, to be sure, dangers to relying on citation analysis —or, for that matter, any one yardstick—to measure a scientist's impact. For one thing, the multiple authorship of most scientific papers muddies the statistics. For another, methods papers tend to be cited out of all proportion to their contribution to basic knowledge, and papers in the life sciences are usually cited more than those in the physical sciences. For still another, some scientists publish mostly in obscure journals not often read. Finally, some papers, as Eugene Garfield has noted, are "so profound in their impact and so quickly absorbed into the mainstream of science" that they no longer need be cited—victims, in this narrow respect, of their own success.

Still, the general relation holds: A few scientists publish a lot, are cited a lot, and make a great impact. If there are Kuhnian revolutions to make, they make them. These constitute the tiny elite corps who, as sociologists Jonathan and Stephen Cole have shown, are named to the prestigious societies, are awarded the honorary degrees and the academy memberships, walk off with the Nobels and the Laskers.

There is one other striking observation that can be made about these elite scientists: They have usually served in the labs of other elite scientists—just as they, in turn, become mentors to the next generation of the elite.

•

The powerful role of the mentor relationship in grooming the scientific elite has been noted by every student of science as a social phenomenon. "There are few scientists of note who did not have an identifiable sponsor," observes Jonathan Cole in *Fair Science: Women in the Scientific Community*.

> Indeed, autobiographical accounts almost invariably contain homage paid to a sponsor relationship, even if it was not always devoid of ambivalence. Whether we examine Auguste Comte's relationship with Saint-Simon; Fermi's with Corbino; Segre's with Fermi; Otto Hahn's with Rutherford; Lisa Meitner's with Otto Hahn; Schwinger's with I. I. Rabi; J. D. Watson's with Luria; Mary Whiton Calkins's with William James; or the thousands of other master-apprentice relationships, we are dealing with one important mechanism of transmitting a scientific tradition from one generation to another.

The mentor relationship's role at the top reaches of science was nowhere better demonstrated than in *Scientific Elite*, Harriet Zuckerman's study of American Nobel laureates. Zuckerman studied the ninety-two Nobel laureates who had done their prizewinning research in the United States by 1972, in the fields of physics, medicine, and chemistry. She found that more than half—forty-eight of them—had worked as students, postdocs, or junior colleagues of older Nobel laureates.

So striking was the phenomenon that Zuckerman devoted several pages of her book to what looked like genealogical charts—names, layered by "generation," with lines connecting them. Except, these were scientific genealogies, showing the progression of scientific influence down through the Nobel-winning generations. In physics and physical chemistry, Glaser had worked in Anderson's lab, and Anderson had worked with Millikan, who worked for Nernst. . . . In physics, it was Bohr and Bethe, both of whom had worked with Rutherford, who had worked for J. J. Thomson, who was a student of Rayleigh. . . . In the biological sciences, Khorana worked for Kornberg, who'd been a student of Carl and Gerta Cori, who worked for Otto Loewi, and so on.

Nor did the phenomenon take hold only with the first Nobel Prizes. "Consider," wrote Zuckerman,

> the German-born English laureate Hans Krebs, who traces his scientific lineage back through his master, the 1931 laureate Otto Warburg. Warburg had studied with Emil Fisher, recipient of a prize in 1902 at the age of fifty, three years before it was awarded to *his* teacher, Adolf von Baeyer, at age seventy. This lineage of four Nobel masters and apprentices has its own pre-Nobelian antecedents. Von Baeyer had been the apprentice of F. A. Kekulé, whose ideas of structural formulae revolutionized organic chemistry. . . . Kekulé himself had been trained by the great organic chemist Justus von Liebig (1803–1873), who had studied at the Sorbonne with the master J. L. Gay-Lussac (1778–1850), himself once apprenticed to Claude Louis Berthollet (1748–1822).

Berthollet, in turn, helped found the *Ecole Polytechnique*, was scientific advisor to Napoleon, and worked with Lavoisier to revise the chemical nomenclature system.

It is through the mentor relationship, then, that elite science—seen as an entity unto itself distinct from everyday or "normal" science—propagates itself. By this view, a great scientific discovery is the product not of individual genius alone but of a scientific "family," down through the generations of which something special, something pivotal, has been passed on.

But what, precisely, gets passed on? Certainly not just specific knowledge and technique; indeed, these may be the least of it. In a long chapter in *Scientific Elite* devoted to "Masters and Apprentices," Harriet Zuckerman noted that it wasn't knowledge or skills that apprentices acquired from their masters so much as a "style of thinking," as one laureate in chemistry told her. It was problem-*finding* as much as problem-solving. Those future Nobel laureates were being socialized, to use sociology's vocabulary, into a sense of the significant, or important, or right problem.

Thus, Zuckerman writes, "aspects of scientific taste are transmitted along chains of masters and apprentices, aided by the apprentices' strong identification with their teachers (sometimes involving what is, for them, a thorough hero worship)." Later, taking on the role of mentor themselves, "elite scientists tend to reproduce in their own attitudes and behavior some of the same patterns they witnessed when they were apprentices."

One might think to liken a mentor chain to the classic parlor game in which one person tells a story to the next in line, who tells it to the next, and so on, the story at the end bearing but flimsy resemblance to the original. But the mentor chain running from Brodie to Pert, at least, is not like that at all. Here, the story, as it were, has been handed down with remarkable fidelity, the scientific legacy of James Shannon and Steve Brodie reaching down across the generations almost intact.

Don't bother with the routine scientific problems, it might read. Leave them to others. Don't bother, either, with big, fundamental problems that are simply not approachable with available techniques and knowledge; why beat your head against the wall? Half the battle is asking the right question at the right time—when it's neither premature to tackle it, nor invites too obvious an answer, when the right methodology is at hand, when enthusiasm is at its peak.

And then, just *do* it. Don't spend all year in the library getting ready to do it. Don't wait until you've gotten all the boring little preparatory experiments out of the way. Don't worry about scientific controls, except the most rudimentary. Just go with your hunch, your scientific intuition, and isolate that simple, elegant, pointed experiment that will tell you in a flash whether you're on the right track. Or, as Steve Brodie might say: Just go ahead and take a flier on it.

One or another link in the mentor chain might well insist on various departures from the script. But all have plainly taken its essence to heart.

Terry Moody, of Candace Pert: "She's always willing to take the long shot."

Julius Axelrod, on his own scientific style: "Do an apparently simple experiment that gives you an important bit of information. . . . Ask the important question at the right time. Ask later and it's obvious."

Agu Pert, on Candace Pert's experimental approach: "Why spend days and days grinding it out, when if you've clever you can do a simple experiment?"

Sol Snyder: "I'd rather do an experiment in three hours than three days or three months."

Axelrod: "You don't learn anything by thinking about what to do, just by going into the lab and doing it."

Robert Goodman, of Snyder: "Sol's attitude is, *Do the experiment. Find out.*"

John Burns on the "gene" Steve Brodie passed down: "Why waste time on uninteresting problems, when there are so many interesting ones to do?"

Axelrod: "It takes about the same amount of effort to work on an important problem as a trivial and pedestrian one."

Sol Snyder, on the standard he tries to apply to every experiment: "Will it win the Nobel Prize? . . . A student will say, 'But it's good science, isn't it?' And I'll say, 'Yes, but it's boring. I think we can do something more exciting.' "

Donald Brown, on how rigorously Axelrod performed experiments: "Rigorously enough."

Snyder: " 'Take it easy,' I'll say to a student. 'Let's not spend a million years trying to prove that two and two is four.' "

Candace Pert, of Snyder's style: " 'Don't think about it,' he'll say. 'Just get hysterical and do it.' "

Not surprisingly, some elements of the Brodie legacy—*not* all—are similar to those Harriet Zuckerman identified among elite scientists generally. She found, for example, repeated emphasis on pursuing the important problem, and at just the right time. "An excess of concern with precision," on the other hand, was rarely encouraged. One laureate told her, about his mentor: "He led me to look wherever possible for important things rather than to work on endless detail or to do work just to improve accuracy."

Mentor relationships, writes Jonathan Cole in *Fair Science*, are "essential in producing in young scientists a sense for a good question or a key problem, a style of doing research or theorizing, a critical stance, and a way of teaching their own future intellectual progeny." In this sense, then, they pass down a store of "secret" knowledge.

•

But what of those not privy to their secrets?

Martin Zatz, a veteran of Julius Axelrod's lab and a scientist of uncommonly broad cast of mind, was talking about mentoring and its role in science. "Are you going to talk about the *disadvantage* of the mentor chain?" he asked me, leaning back in his swivel chair, hands clasped behind his head, smiling broadly.

What's that? "That you don't get anywhere," he replied, now quite serious, "unless you're in one."

Lacking an influential mentor, Zatz was saying, one is almost inevitably excluded from the ranks of the scientific elite. He was stating a corollary, as it were, of the Matthew Effect.

In 1967, sociologist Robert Merton (Harriet Zuckerman's mentor at Columbia University, by the way) went before an American Sociological Association meeting in San Francisco to deliver a paper, later to appear in *Science*, that would instantly enrich the sociological literature. It was called "The Matthew Effect in Science," and it took its name from the Gospel According to Saint Matthew, where it is written, "For unto every one that hath shall be given and he shall have abundance. But from him that hath not shall be taken away even that which he hath." In other words, Them that has, gets.

Merton described what he saw as "the accruing of greater increments of recognition in particular scientific contributions to scientists of considerable repute and the withholding of such recognition from scientists who have not yet made their mark." Two scientists make a simultaneous discovery? The more established is apt to get the credit. Several scientists collaborate on a paper? The best known among them is seen as the brains behind it.

Other sociologists have extended Merton's model, and the now widely held view sees science as not only sharply stratified into haves and have-nots, but with Matthew Effect phenomena conspiring to keep it that way: Bright, ambitious Ivy League grad gains admission to top graduate program . . . lands postdoc with eminent researcher . . . latest equipment, plus competitive environment, plus distinguished mentor, plus important problem, leads to noteworthy papers in top journals . . . famous mentor's contacts grease the way to solid position at fine university . . . gets best students, who contribute to discoveries, which solidifies reputation (along with that of own mentor as well).

The scientifically rich, to put it another way, get richer.

Half of all the Nobel laureates in Harriet Zuckerman's sample received their doctorates from just four institutions—Harvard, Columbia, Berkeley, and Princeton. At the time of her study, almost three quarters of the 710 members of the National Academy of Science who'd earned American doctorates had done so from just ten universities. Then there was the astounding frequency, already cited, with which Nobel laureates counted other laureates as their mentors.

As Martin Zatz implied, failure to hitch onto the right mentor exerts a powerful drag on one's career. The master and apprentice system is anything but egalitarian, sometimes magnifying real differences in ability, sometimes unduly rewarding those who happen to be in the right place at the right time.

On the other hand, it is hard to fault a system that, on the whole, so well identifies and rewards talent. In a world of cruelties and injustices far more odious than any in science, does the mentor system's accumulation of advantage in a few seem so high a price to pay? If differences in ability are magnified, are they not at least there to be magnified in the first place? If scientific success breeds further success, is not the initial success, at least, well earned? Has not the Brodie chain left generations of pharmacologists imbued with a rich tradition of research excellence, opened up whole new scientific disciplines, been responsible for crucial discoveries, led to the development of valuable new drugs, and, all in all, done no inconsiderable amount of good?

One long-time observer of science, syndicated columnist Daniel Greenberg, recently devoted an essay to what he called "the serfs of science"—the graduate students and postdocs who do most of the actual bench work. "The elders of the profession exploit these aspiring youngsters," he wrote. Senior authors, he noted, often take credit for work done, and even conceived, by junior people. Yet in the end, he conceded, while "the system may be outrageous, . . . it surely is productive. And the victims can be confident that after they make the grade, a new generation will be clamoring for entry."

Still, in the relationship between mentor and "victim" lies fertile ground for resentment and emotional strife. "Even a cursory scan of the autobiographical histories of scientists," Jonathan Cole noted in *Fair Science*, "reveals poignantly how sponsorships involve jealousies, ambivalence, [and] conflict," in addition to admiration and love. Harriet Zuckerman found much the same among the Nobelists she studied. Apprentices got too much attention from their masters, or too little. Their masters expected too much from them or showed insufficient gratitude for what they did do. Joint work credited to the master sometimes became a source of outright conflict.

Rarely do mentor relationships progress with such harmony as apparently marked that between Sol Snyder and Julius Axelrod. Axelrod's resentment for Brodie, and Pert, for Snyder, are more typical, even predictable, given the intensity of their relationships.

A case can be made for recording only the contributions of a Brodie, an Axelrod, a Snyder, or a Pert, and not bothering with whatever animosities might surface among them. Is it not their formidable intellects, their scientific achievements that count, and that alone? Is it not enough to record that Bernard B. Brodie was a great and honored pharmacologist and that in his laboratory, one day in the early 1950s, the enzymes God or nature created for the detoxification of foreign substances were first discovered?

Just such a case is made by Elliot Vesell, Steve Brodie's

former student and distant cousin. Vesell sees science as a lofty enterprise, insists that only its grandeur and greatness should be recorded in print. For him, Brodie is an exemplar of all that's best in science, living out the dedication to truth represented by Sinclair Lewis's idealistic researcher, Arrowsmith, in his novel of that name.

Yes, perhaps other parts of the personality of a Steve Brodie, say, are not unfailingly attractive. Perhaps rivalries and hard feelings do surface in the life of such a luminary. But these, says Vesell, are irrelevant and have no legitimate business in any account of the man and his work. For him, he repeats, it is only that part of Brodie in the Arrowsmith tradition that warrants attention.

Except that Martin Arrowsmith never lived; he was the product of a novelist's imagination. Steve Brodie, mercifully, is not; today he lives in a low, flat desert house on a quiet, sun-baked street in Tucson, Arizona.

And eight hundred miles to the northwest, in rainy Portland, Oregon, lives his old friend, the tall, bespectacled Irishman with the neat bow tie in whose laboratory he first blossomed into a scientist—James Shannon.

13.
1985

HE IS PAST EIGHTY NOW, in good health, and living in a house on Homewood Street in Portland, Oregon. His daughter Alice, a practicing physician, lives nearby. He has been retired from the directorship of the National Institutes of Health for more than fifteen years.

Back in the late 1960s, James Shannon's anticipated retirement—which, as a Public Health Service officer, was set at age sixty-four—provoked both tributes to his tenure and alarms at the prospect of his departure. "His loss," Irvin H. Page told the readers of *Modern Medicine*, "will be so great that each of us must take on some of the responsibility of seeing that a worthy successor is chosen." *Science* observed that, with Shannon retiring, "the last of the postwar giants will be gone. . . . To the extent that any major government enterprise can be considered the work of one man, the billion-dollar-a-year National Institutes of Health is the work of James A. Shannon."

In the mid-1960s, for almost the first time, NIH had begun to suffer criticism. Some said it focused too much on basic science and not enough on the care of the sick. Others worried how all those health research millions were being spent.

In 1965, a presidential commission set out to investigate Shannon's biomedical empire. Teams of scientists, administrators, and consultants visited thirty-seven institutions then receiving NIH grants as well as the labs of fifty-two intramural researchers, interviewing not only hundreds of those awarded NIH grants but many denied them.

Conclusion? "The activities of the National Institutes of Health are essentially sound and . . . its budget of approximately one billion dollars a year is, on the whole, being spent wisely and well in the public interest." The area of greatest concern? That certain organizational and procedural weaknesses portended future problems; NIH's success was too dependent on "the unusual qualities of a few individuals"—by which the commission meant Shannon and his hand-picked top staff.

As the time for Shannon's retirement neared, some made moves to stop it. One editorial cartoon showed a bow-tied Shannon, bag of golf clubs slung over his shoulder, pockets stuffed with travel and retirement village brochures, bounding from his NIH office as two arms restrained him—those of then-Secretary of Health, Education, and Welfare Wilbur J. Cohen and President Lyndon Johnson. The accompanying editorial was entitled, "The Indispensable Man." Shannon, "the agency's guiding genius," it said, should be kept at his post for at least two more years.

But in September 1968, right on schedule, Shannon did retire—by some accounts hurt that, in the end, no way had been found to waive his mandatory retirement, as was done for the FBI's J. Edgar Hoover and Navy Admiral Hyman Rickover, who both served well past normal retirement age.

Shannon became a special advisor to the president, took a post with Rockefeller University in New York, went on to a succession of consultancies and boards with various academic and medical institutions, received numerous honorary degrees—eight in the two years following his retirement alone—and continued to voice his views on issues of science and public policy.

He returned to Bethesda in 1976 for the unveiling of a bronze bust in his honor. "The podium is yours," Dewitt Stetten, then NIH's deputy director, said in introducing him. "You may tell us how to conduct the next twenty-five years." Shannon may have wished to do just that. "The old gang gathered in the director's office," Stetten remembers, and confided in their old boss their current problems. " 'I'll tell what you ought to do,' " Shannon would say—and did. "He had never surrendered the job of director," says Stetten. Jack Orloff, too, remembers Shannon, long past retirement, sometimes calling him at NIH, bemoaning one action or another of his successors. "You don't think I want to throw away fifteen years of my life," Shannon once grumbled to him in frustration.

On September 29, 1979, Shannon and some thirty other Goldwater oldtimers gathered at the Embassy Row Hotel in Washington, D.C., for a reunion. "It was very nostalgic, warm, and pleasant," says Shannon's old friend Thomas Kennedy. "Everybody had a wonderful time." Robert Berliner was there. So were Sidney Udenfriend, and Steve Brodie, and Julius Axelrod. Shannon was presented a glass bowl, etched with the names of those present.

"Oh, he enjoyed it," says Kennedy. "He was amazed that people would go to all the trouble." He was more amazed yet —"flabbergasted" as Kennedy puts it—when, two years later, NIH's colonnaded Building 1, from which he'd overseen the growth of NIH for a decade and a half, was renamed the James A. Shannon Building, as five hundred friends and former colleagues looked on.

Each participant in the reunion had been asked to prepare an update on his or her activities since leaving Goldwater, most taking the opportunity to address warm words of thanks to Shannon. Shannon himself, seventy-five, succinctly recounted the highlights of his career, concluding with the information that he'd recently bought a house in Portland. "I might add that overall my journey has been pleasant and stimulating," he wrote in his spare, scratchy hand to David

"Bud" Earle, reunion organizer and Shannon's successor at Goldwater, "not the least reason for this being the exciting people it has been my pleasure to know and to work with.

"And so, Bud, it goes. My plan is a rolling plan and program developed in anticipation of an annually extending five-year segment, each year being the first year of a five-year cycle during which time I hope for reasonable health and continuing companionship in a quite busy life."

•

Goldwater Memorial Hospital still serves as a busy chronic disease care facility, but its feeling of freshness and architectural novelty has been dimmed by the passage of years. Welfare Island, at whose southern end it sits, is called Roosevelt Island now. The hospital now shares possession of it with, at its north end, a much touted "new town" of mid-rise apartment buildings. The trolley from Queens is gone now, the last one having creaked its way to the island on April 7, 1957. Buses replaced it. Today a cable-pulled tramway from East Fifty-ninth Street in Manhattan soars high over the East River to a terminal just north of the hospital. In the basement of Building D, an arrow still points the way to SPECIAL RESEARCH LABORATORIES, DIVISION 3, NEW YORK UNIVERSITY SCHOOL OF MEDICINE UNIT, where forty years ago Shannon, Brodie, and the others waged war on malaria.

Malaria remains a subject of active research interest, though not at Goldwater. Today, the conscientious objectors and prison inmates of the war years have given way to army volunteers and civilians, in some cases paid fifteen hundred dollars to suffer the chills and fever dealt out by the *Plasmodium* parasite. The *Anopheles* mosquito, for a time brought under control by DDT, has developed strains resistant to it. Atabrine has been replaced by more potent drugs. Chloroquine, among other compounds developed in the closing days of the war, was for many years the drug of choice against malaria, but the parasites grew resistant to it, too, in time.

So the search for new drugs continues, one named mefloquine today being among the brightest hopefuls. A vaccine against malaria is also in the works. In 1984, microbiologist Ruth Nussenzweig announced that she and a team of researchers under her direction at New York University had developed a vaccine that protected monkeys against the disease that continues to kill a million African children every year.

•

As the Lasker episode receded into the past, a color photograph of Candace Pert talking animatedly with Jonas Salk, discoverer of the polio vaccine, hung in her office in 3N258 of the NIH Clinical Center. "It reminds me," says Pert, "that I'm more interested in curing disease than in publishing papers."

In 1977, Pert had been promoted from staff fellow to senior staff fellow of the section on Biochemistry and Pharmacology, and in 1978 to research pharmacologist. In 1982, she was named chief of a new Section on Brain Biochemistry of the Clinical Neuroscience Branch. By 1984, she had more than 110 scientific papers to her credit, had won the Arthur S. Fleming Award of the American Chemical Society, had delivered a number of honorary lectures, and was on the editorial board of five scientific journals.

In the years following the Lasker controversy, she had also been the subject of numerous articles in such popular magazines, as *Fortune*, *Redbook*, and *Omni*. Her colleague, Miles Herkenham, saw this as "a good direction for her, as a liaison between science and the media. Her value lies in generating the excitement of science. She can pump people up." Her *Omni* interview was later included in an anthology whose publication was marked by a New York reception for the interviewees. It was there that she was photographed with Jonas Salk.

"I've come into the fold again," she said in 1984, feeling that she was being taken more serious than in the period after the Lasker flap. "It's because I'm such a good scientist. When I'm up there with Jonas Salk, they have to recognize me. I've worked very hard for everything I have."

Her split from husband Agu came abruptly. "One day I

just moved out," he says. "There was a lot of tension from the way she was getting recognition, things I couldn't handle well. A part of me was a bit resentful. It definitely affected our relationship."

Pert, mother of three children ranging in age from teenager to two-year-old, still scientifically collaborated with Agu and their relationship remained cordial. But more often she worked with Michael Ruff, an NIH immunologist seven years her junior. Late in 1984, they published in a European journal evidence that neuropeptides, including a form of enkephalin, and probably bombesin and substance P as well, powerfully stimulate the migration of macrophages, cells originating in the bone marrow that flock to injured tissue and contribute to wound healing.

Their collaboration extended into the personal realm as well, and Pert was as excited about their romantic ties as their professional. "Ruff and Pert" she'd say out loud, just to hear the sound of it. "I have a romantic feeling for scientists. I worship them."

At thirty-eight, Candace Pert still gave herself over freely to her whims and enthusiasms, was still, uniquely, Candace Pert. One recent cool spring day, her toenails were done up in red and black polka dots, her office in ubiquitous rainbows. A set of glass laboratory cannisters filled with brightly colored liquids lined one shelf. Her blue wool pea jacket was thrown across one of five chairs somehow squeezed into the tiny office. A file cabinet jutted out at a crazy angle from the wall. Her Arthur S. Fleming Award hung crookedly above her desk. Scraps of paper littered the floor. A clock on the office wall kept time, of a sort, its second hand lurching ahead a few seconds, hesitating, then again surging forward. It read 5:52. It was 3:05.

•

In 1985, at the age of forty-six, Solomon Snyder stood exactly on the cusp between young and old. One moment the fine wrinkles on his face would seem to fall away, laying bare the clean-featured forms of young adulthood. But then, one had only to imagine his facial contours rounded and softened even just a bit, and in the blink of an eye you could see how he'd likely look at sixty-five.

Snyder was in his prime. In 1980, he was named to the National Academy of Sciences. In 1983, before the Israeli Knesset, he accepted his share of the prestigious hundred-thousand-dollar Wolf Prize, along with Jean Pierre Changeux of France and Sir James Black of England. In 1984, while I sat in his office, he excused himself to take a call and learned he was the recipient of the first Einstein Award for Research in Psychiatry and Related Disciplines from Yeshiva University. In the years before physicist Murray Gell-Mann won the Nobel Prize in 1969, it is said that cocktail party talk among physicists ran to, "I wonder if Murray will get it this year." In 1985, much the same climate surrounded Snyder. (Meanwhile, his scientific father, Julius Axelrod, worried that "Sol is expecting it, and sometimes you can expect it too much.")

Papers continued to pour from Snyder's lab, filling the pages of his professional bibliography, which at forty-eight pages and growing already tested the fastening capacity of the standard office stapler.

Gould, R. J., Murphy, K. M. M. and Snyder, S. H. A simple sensitive radioreceptor assay for calcium antagonist drugs. *Life Sciences*. 33:2665–2672, 1983.

Snyder, S. H. Drug and Neurotransmitter Receptors in the Brain, *Science*, 224:22–31, 1984.

Javitch, J. A., Blaustein, R. O. and Snyder, S. H. ^{3}H-Mazindol Binding Associated with Neuronal Dopamine and Norepinephrine Uptake Sites. *Molecular Pharmacology*, 26:35–44, 1984.

One story heard around Johns Hopkins was that there had to be something wrong with you if, after joining Snyder's lab, you had no paper to show for it in three weeks. A harsher version painted Snyder's lab as so efficiently aimed at scientific production that writing a paper amounted to filling in the data blanks and the authors' names.

Snyder's production extended to more popular genres as

well. He has writen encyclopedia entries, popular magazine articles for the *New York Times Magazine*, and a string of popular books, like *The Uses of Marijuana* and *Madness and the Brain.* In 1984, he was writing an account of the discovery of the opiate receptor, which he described as in the tradition of *The Double Helix*, James D. Watson's gossipy, startlingly frank account of the discovery of the structure of DNA. Snyder was also working on a book about drugs and the brain, for *Scientific American*, which, he'd delight in telling you, was virtually sure to sell forty, maybe fifty thousand copies.

On October 24, 1984, in a ceremony held against the backdrop of a computer-enhanced color image of a radioactive slice of rat brain, a ribbon was cut to mark the opening of the new Baltimore laboratories of Nova Pharmaceuticals. According to its exquisitely produced annual report, Nova was the first company "to utilize as its core business state-of-the-art neuroscience for the discovery of new drugs"—more particularly the receptor technology pioneered in Snyder's lab.

On hand for the dedication were, among others, Bernard L. Berkowitz, head of the Baltimore Economic Development Corporation (BEDCO); Donald G. Stark, Nova's chief executive officer and former president of Sandoz Pharmaceuticals, Ltd. of Japan; and, dressed in a smart blue suit, smiling broadly, Solomon H. Snyder, chairman of Nova's scientific advisory board, and both a consultant to, and a director of, the fledgling company.

It was a big day for BEDCO, a chance to trumpet its success in bringing a shimmer of high tech to gritty, blue-collar Baltimore. A shuttle bus ferried members of the media from a luncheon at the Maryland Science Center at the edge of the city's gleaming new Inner Harbor to the dedication ceremony on the grounds of Francis Scott Key Medical Center in East Baltimore. There, Nova president Stark acclaimed the transformation—"from rubble to research," he said—of an abandoned, debris-strewn kitchen building into modern laboratories.

Though by early 1985, Nova had not yet gone into production, it had reached agreements with Johns Hopkins University for marketing rights to patent-pending new compounds called arylxanthines, for a radioreceptor assay for calcium channel blocking drugs, and for nausea preventatives potentially useful in cancer chemotherapy. A six-million-dollar offering of Nova stock in July 1983 had made Snyder, at least on paper, a rich man.

•

In January 1970, a group of Julius Axelrod's former students, sparked by Sol Snyder, met at a scientific conference in Paris to discuss a *festschrift* for their mentor. A *festschrift* is a volume of essays or articles, contributed by colleagues and admirers, bestowed as a tribute. They would present it to Axelrod, they decided, at the annual meeting of the Federation of Societies of Experimental Biology the following spring.

In November, they got scooped; Stockholm saw fit to honor Axelrod first. But plans for the *festschrift* went ahead on schedule. As Seymour Kety wrote in an introduction to the volume, "the action of the Nobel committee simply confirmed the conviction that we were on the right track!" The *festschrift*, which comprised a dozen or so specially written scientific articles by his former students, was published by Oxford University Press in 1972 as *Perspectives in Neuropharmacology: A Tribute to Julius Axelrod*, it having been first presented to him the year before in Chicago. A framed collection of photos of the twenty-three of "Julie's boys" participating has hung behind his desk since.

When Axelrod won the Nobel in 1970, he wondered, "How will it change my life? What will it mean to me?" With his twenty-three-thousand-dollar piece of the prize money, he bought a stereo and furniture, gave some to his kids, and invested some in stocks and bonds. "Now if you need money, I can loan it to you," a friend remembers him saying.

Inwardly, the prize pleased him deeply. "When you look back, it makes you feel good," he says, "like when you get depressed." Outwardly, his life changed but little. He was, of course, showered with the usual run of post-Nobel honors, including three honorary degrees and eight honorary lectureships the next year alone. And there were the incessant pleas to speak out on one political issue or another, to sign this or that manifesto. "Some I did sign," he says.

But he continued to live in the same high-rise apartment building at 10401 Grosvenor Place in Rockville, Maryland. He still often took the bus to work. He remained accessible to his students. One of them, Michael Brownstein, remembers Axelrod mentioning the Nobel exactly once during the whole following year. Brownstein was applying for a bank loan and needed a letter testifying to his employment. He went to Julie. "Would it help," asked Axelrod, "if I sign it, 'Julie Axelrod, Nobel laureate'?"

The Nobel did bring one major change, however. At the recognition ceremony held for him by NIH, Axelrod expressed hope "that more Nobel Prizes will be won by NIH and NIMH scientists. And I hope somebody will win one pretty soon so I'm forgotten and I can get back to the lab again." He did get back to the lab again, but it was never quite the same.

In *Scientific Elite*, Harriet Zuckerman reported that the period after winning the Nobel was a potentially difficult one for the new laureate, his research output dropping, on average, by about a third during the following five years. Axelrod escaped that curse; in 1970, he published seventeen papers; in 1971, twenty-one papers; in 1972, also twenty-one; in 1973, twenty. But there was one difference: He was no longer doing the work with his own hands; at the age of fifty-eight, he was giving up the two or three experiments per week he'd been doing for years. So long as he remained "intimately involved" with the bench work, he reasoned, it would be all right.

As Axelrod approached seventy, he wrote a number of more

personal accounts of various aspects of his life's work, usually with more approachable titles, like "My On and Off Research on the Pineal Gland." One subject he reviewed was microsomal enzymes. Writing in *Trends in Pharmacological Sciences* in 1982—more than a quarter-century after the work itself, more than a decade after having won the Nobel Prize, the holder of almost three dozen other important awards, medals, and honorary degrees, his reputation never more secure—Axelrod still had something to get off his chest. He did so, in a paper called "The Discovery of the Microsomal Drug-metabolizing Enzymes," which set forth his version of their discovery.

Early in the 1980s, Axelrod was still busy doing research in, and publishing regularly on, such areas as cell membrane function, pituitary secretion, and lipid chemistry. He still kept a postdoc and two others busy with ideas, and followed closely the careers of his former students, for whom, it seemed, he was forever writing letters of recommendation. "Having worked in this lab opens many doors," he says. "I feel great when my students do well. But they always do well."

All through these years, Axelrod could be found at his desk in 2D45. There he is now, tieless, in a black-checked, short-sleeved sport shirt, shapeless dark pants, and comfortable-looking suede shoes. He makes his way noiselessly down the long Building 10 hallways, left arm tucked in close to his body, left hand in pants pocket, with a shuffling ease that speaks of long years avoiding corridor collisions. It's been thirty-five years since he first saw the brief article in the *New York Times* about James Shannon being appointed research head of the Heart Institute.

In 1984, Axelrod retired. Except that Axelrod's version of retirement changed nothing but NIMH personnel records. Formally, he became a guest scientist in the NIMH laboratory of his old student, Michael Brownstein, moving a half-mile up the hill to Building 36. Retirement allowed him to do outside consulting, and to sit on boards and be paid for it, something for which, as a government employee, he was ineligible. NIMH, meanwhile, was able to free up a position. So far as his research was concerned, nothing changed. He still had his own students, was free to follow his scientific nose.

It was an arrangement novel and newsworthy enough to merit a page in *Science* under the head "Retiring Frees NIH 'Guest' to Consult." Axelrod was quoted as saying it would be "like being a gentleman scientist in the old days. I have the best of two worlds."

Among the companies for whom retirement freed him to consult was Nova Pharmaceuticals, Sol Snyder's company, on whose scientific board he now sits and from the pages of whose annual report his smiling countenance stares benignly. Snyder also helped secure funding for Axelrod's post-retirement research by reminding Johnson and Johnson/McNeil, the drug company whose corporate coffers Tylenol has helped fill, of his mentor's role in its discovery. The big pharmaceuticals house came through with a twenty-thousand-dollar contribution to Axelrod's research budget.

In May 1984, a scientific symposium was held at NIH to mark Axelrod's "retirement": Two days of scientific talks, twenty in all, under the title of "Mechanisms of Synaptic Regulation," followed, on the evening of the second day, by a lavish testimonial dinner. Word was spread around NIH and elsewhere through large, pastel-colored posters depicting a stylized, free-form synapse, the "Synaptic" of the title being executed in a rakish, blood red script designed just for the occasion.

"If you look through the program, it's a Who's Who of the neurosciences and neuropharmacology in this country and around the world," NIMH scientific director Fred Goodwin said in opening the symposium. All the speakers were former Axelrod students. Each got up, expressed a few words of thanks or acknowledgment to Julie, told an anecdote; sometimes, before launching into the substance of their talks, they'd flash on the overhead screen the title page of a paper on which they'd collaborated with him years before.

Many, throughout the two days, referred to Julie's scientific "children," "grandchildren," or "family." NIMH researcher Terry Reisine, for example, told how he'd first worked for Henry Yamamura, Axelrod's scientific grandchild through Sol Snyder; how when he'd visited Jacques Glowinski in Paris, he'd seen Axelrod's picture in Glowinski's office; how, arriving in Axelrod's lab, he'd in turn found Glowinski's picture. "I'm proud," he concluded, "to be part of the family."

Axelrod's "children" were grown now. Looking out from the *festschrift* collage or from other photos on display later at the banquet, their skin was smooth, their youthful faces earnest, their hair short and slicked back, or else long and tousled, depending on the decade of their apprenticeship. There was Joseph Coyle in wide-wale corduroys, fisherman's sweater, and sideburns down to his mouth. And Glowinski with a pipe stuck out of a corner of his mouth, very much the suave European intellectual. And Snyder, looking uncharacteristically stern and serious. Now here they were back in Bethesda to honor their mentor, all major names in their own right now, older, their faces filled out, their bodies and their bearings substantial with responsibility.

At the retirement banquet where Jacques Glowinski wrote his "book," held the evening of the second day, a single peach-colored rose graced each table, a five-hundred-milliliter laboratory beaker serving as a vase. The long head table, decked out with pink table cloth, and well-stocked with dignitaries, stretched across the hall of the Chevy Chase Women's Club. Among those on hand were Sally Axelrod, sitting up ramrod straight at the head table, jaw tight, mouth set; Candace Pert, in a black, shoulderless dress, hair piled on top of her head in curls; and Sol Snyder, looking dapper in a tuxedo he'd gotten stuck wearing by a last-minute change of plans about which he'd been notified too late.

The menu reflected Axelrod's fondness for things French: Tenderloin of beef with sauce perigourdine, asparagus vinaigrette, croissants, miniature French pastries, French roast coffee. The wine was Pouilly-Fuisse 1982, and two other French varieties. A guitarist and a flutist played softly before and during dinner. Poster-sized photos of the guest of honor were scattered around the periphery of the hall. One, looking like an old publicity shot, obviously posed, showed him pipetting something at the lab bench. Another showed him, amidst the gowns and jewels, tuxedos and medals of Swedish aristocracy, receiving the Nobel Prize.

At one point during the two days, Candace Pert went up to him, hugged him, and asked, "Aren't you embarrassed by all this adulation?" "No," Axelrod whispered, "I love it." During the dinner itself he got up to say how he was "pleasantly embarrassed" by the tributes, and felt *nachas*—the Yiddish word for mingled pride and pleasure—at seeing all his students there.

As to the future, he said, "I have no plans to retire." With two scientific lives behind him, he was looking forward to a third.

Two months later, an occasional collaborator of Axelrod's, Merrily Poth, was sharing her excitement over a piece of his recent work. "It's creative, original, important—really *pretty*." He had a postdoc helping him, she said, "but I know it's Julie's idea, I just know it. He's seventy-three, for God's sake. He shouldn't *do* that. He should let *us* have something!"

•

Steve Brodie retired in 1971, scared into it by two heart attacks and the advice of his doctors.

Brodie's medical problems went way back. In the 1960s there were many times when he ran the lab from the bedroom of his apartment. One time, plagued by pain, jaundice, and blood in his stool, he was sure he had cancer. His friend, Mimo Costa, shepherded him to New York, where he was treated for a gall bladder problem, a prolapsed colon, and a variety of other ills. He was in the hospital when he heard about winning the Lasker Award. The award ceremony itself took place between two bouts of surgery.

In 1972, a big dinner was held for him in San Francisco, coordinated with that year's international pharmacology conference. It wasn't billed as a retirement dinner; "Everybody will think I'm dead," Brodie worried. But that's what it was. "The whole thing was engineered beautifully," remembers Victor Cohn, now professor of pharmacology at George Washington University and one of those there that day. The whole world of pharmacology was out for it—the top names from the United States, Europe, and Asia, from industry, from academia, many of them with their spouses. Says Cohn, "It's hard to imagine a more stellar group."

The affair was held in a large hall on whose periphery were tables for eight or ten people each, with room for dancing in the middle. At the far end of the room as you came in was the head table. Off to the right were two thick books, bound in rich black leather, with *Bernard B. Brodie* stamped in gold on the covers, the first page elegantly and flawlessly calligraphed:

PRESENTED TO

BERNARD B. "STEVE" BRODIE

by his

PRESENT AND FORMER COLLEAGUES

SAN FRANCISCO, CALIFORNIA

JULY 26, 1972

The two books, together set in a black leather-covered box, were filled with warm letters of appreciation from some of the world's most prominent pharmacologists. They'd come from Bern, Switzerland, from Dublin, Ireland, from Mainz, West Germany; from Milan and Bethesda and Los Angeles and Taipei and New York and Prague; and from Minsk, in the Soviet Union—all Brodie's scientific children, testifying to the mark he'd made on their lives, each expressing thanks, recounting a lesson learned, or recalling a moment shared.

"Dear Dr. Brodie: How do I express my gratitude for being given the privilege of working with you these many years? . . ."

"Dear Dr. Brodie: You, John Burns, and I have one thing in common—at one time we were possessors of that little office at Goldwater right next to the cold room. There was hardly enough room for a desk, two chairs, and a file cabinet. . . ."

"Dear Steve: You were the research king of the [Heart Institute] during our golden era. . . ."

"Dear Dr. Brodie: Your flexible enthusiasm for the daring, the innovative, even the controversial permeated those around you. . . ."

"Dear Dr. Brodie: I shall always remember our early-morning sessions from two until five A.M. . . ."

"Dear Dr. Brodie: I vividly recall those periodic meetings held in that small, dingy room [at Goldwater] which served as a library for the Research Service. . . . In retrospect, I now realize that chemical pharmacology, as we presently know it, was being developed [there]. . . ."

"Dear Steve: It has now been six years since I left the NIH, and there have been countless times when I have found myself saying to my students the very same things that you used to say to me. . . ."

"Dear Steve: You literally pointed the way for my entire professional career, and I have never once regretted taking that path. It has been a truly rewarding life, and I thank you for introducing me to its pleasures. . . ."

"Dear Steve," wrote Julius Axelrod, "It was about twenty-five years ago that I first visited you to ask advice about a problem. . . ."

It was a grand bash, with "good food, good drink, good music, good dancing, the kind nobody ever wanted to leave," says Victor Cohn. "There was a great feeling of camaraderie. It was like a family reunion with everyone feeling beautiful about everybody else."

•

The affair had been organized by Hoffmann-La Roche, the drug company, cohosted by John Burns, its director of research and development, and Sidney Udenfriend, director of the Roche Institute of Molecular Biology. Roche amounts to "a mini-Brodie lab," as Victor Cohn puts it, with many of its ranking researchers Brodie disciples. In particular, the Roche Institute originated, its literature says, with "three scientists who were all members of the same NIH laboratory." The laboratory in question, of course, was Brodie's.

In 1967, Brodie had given a party for Jim Shannon at his Bethesda apartment. At one point, as Udenfriend reconstructs it, he and Burns, who'd left Brodie's lab for private industry in 1960, got to talking about the primitive state of the biological sciences in the pharmaceuticals industry. What was needed, said Udenfriend, was a kind of Bell Labs of the life sciences—a reference to the prestigious private research arm of American Telephone and Telegraph.

He thought no more of it until, a week later, he got a call from Burns, following up on the idea. Pretty soon, the giant Swiss-based pharmaceuticals house of Hoffmann-La Roche had set up the Roche Institute of Molecular Biology, dedicated, in the words of its charter, "to fundamental research in biochemistry, genetics, biophysics, and other areas in the domain of molecular biology." Sid Udenfriend was named its founding director—which he remained until 1983, when he passed the reins to Herbert Weissbach, another Brodie student.

To staff the fledgling institute, Udenfriend brought twenty people, many of them Brodie veterans, from his lab at NIH. Roche Institute, he says, was conceived as "the best of NIH, industry, and the university." He feels it's fulfilled its promise, being home, as he'll tell you, to more members of the National Academy of Sciences than, for example, any medical school in New York City. Its thirty senior scientists and seventy postdoctoral fellows are housed in a modern five-story laboratory building in Nutley, New Jersey. Its board of scientific advisors is packed with Nobel laureates.

It is, as he says, "another spin-off from the Goldwater."

•

Following Brodie's retirement, he and Anne lived first in Arizona, then in Palm Beach, Florida. (Anne felt strongly, says Mimo Costa, that "nobody should retire where one once was great.") In Palm Beach, they were located about a mile and a half from the Kennedy family compound, the area was littered with millionaires, and soon Brodie grew bored by conversations that inevitably come down to how you'd made your money. Still, he loved their house there, with its veranda, its flowers and grapefruit trees, its access to the beach. He remained in close touch with scientific developments, was on the phone a lot, and kept busy reading and thinking.

Two or three times a year, at least until his health deteriorated, Brodie would travel up to Roche, for which he was a consultant, usually stopping in Chevy Chase to visit Costa. He'd also frequently stop in Hershey, Pennsylvania, where he was, in retirement, listed as professor of pharmacology at the Penn State medical school there. People "would be kind of shook" after he left, remembers Elliot Vesell, the school's chairman of pharmacology. "They were amazed at his penetration and brilliance."

Ultimately, the Brodies returned to Arizona. He'd always been fascinated by the place, says Costa, whose office at Saint Elizabeth's Hospital in Washington, D.C., displays a large portrait of his mentor. "He was very interested in desert animals, like the desert rat, thinking to use it as a model in biological experiments." At one point, Costa's son—the one for whom, as a boy, Brodie had mischieviously doctored an audio tape of a baseball game—was looking for someone with whom to do his Ph.D. dissertation. "What's wrong with me?" Brodie asked indignantly when Costa mentioned it. Soon Costa's son, Max, was under his wing at the University of Arizona in Tucson, to whose pharmacology department he maintained ties.

When Brodie's health problems left him unable to continue in that role, Diane Russell took over. Diane Russell was Sol Snyder's student, having moved to a position with the University of Arizona following her stint with the Cancer Research Institute in Baltimore. From Snyder, she'd heard the whole soap opera that was the Brodie and Axelrod story. Now, in Tucson, she was thrown up close with the legendary figure of Brodie himself. In 1976, the generations collided in a paper appearing in the *Proceedings of the National Academy of Science*, called "Activation of 3':5'-Cyclic AMP-dependent Protein Kinase and Induction of Ornithine Decarboxylase as Early Events in Induction of Mixed-function Oxygenases." Its five authors included Max Costa, Bernard B. Brodie, and Diane H. Russell.

Though meeting him while already well along in her career, Brodie had "an incredible influence on my thought processes," Russell says. He was long past retirement, in frail health, and close to seventy. Yet intellectually she at first felt "ineffective" with him; he challenged everything she thought she knew. He had a way of reducing a scientific discussion to an ever "simpler, more basic level. 'If this is true, then what?' 'If that's true you'll have to show that . . .' You think, 'Could it be he's right?' "

He was not entirely popular among his new, occasional colleagues at Arizona. At one point, he had a run-in with the head of the department who, Russell says, "felt threatened" by the famous pharmacologist from back east. Around the university, "it became unpopular to really like him." She, though, came to admire and respect him more and more, and today, in her University of Arizona office, a wood-framed formal portrait of him dominates one wall.

Russell also became close with Anne Brodie, often going to her for advice when she had a personal problem; the older woman, says Russell, "accepts me as a daughter, and as a daughter I can, in her eyes, do no wrong." For Mrs. Brodie, reports Russell, "Axelrod is a dirty word. She feels he was not properly grateful" for all her husband did for him. Russell attributes her resentment—which she's never heard expressed by Brodie himself—"to the fact that Brodie wanted the Nobel Prize so much and that Julie got it."

Some think Brodie could still get the prize. "You never know," says Elliot Vesell. Didn't Barbara McClintock win it in medicine when she was eighty-one? While "unlikely," he admits, "it's not out of the question." Brodie could get it in the area of drug metabolizing systems and their relationship to cancer. Or perhaps for his work using reserpine as a tool to probe neuronal function. "Brodie's name has been put in on many, many occasions," says Vesell, Julius Axelrod being among those reportedly submitting it.

For a while, in 1984, the Brodies considered moving to Hawaii, even made an exploratory visit there. But early in their trip, Anne Brodie slipped, fell, and badly bruised her arm, throwing a damper on the trip and canceling their plans.

Today the couple live in a low, brown stucco house on East Mabel Street in Tucson, located a few blocks from the University of Arizona Health Sciences Center, where Brodie has ready access to the medical care he'd need in an emergency. Cacti and other desert plants grace a grassless front yard under the Arizona sun.

One room of the house is furnished as Brodie's study. His red leather chair sits before his old desk, the pair of black leather books given him at the retirement dinner sitting just behind it on a shelf. One wall is a bookcase stocked with scientific texts. The others are lined with his awards and honorary degrees, dozens of them, taking up every available square inch, his name and face looking down from everywhere. There are brass plaques, certificates inscribed with his name, and photographs of him being conferred honorary degrees. Atop a display case are awards that cannot be hung, boxed plaques and medals and statuettes, including the Winged Victory of Samothrace, the Albert Lasker Award.

Hung above one stretch of bookcase is a rectangular red pennant with Chinese ideographs in gold, given him by a Taiwanese student, C. C. Chang. The ideographs represent PIONEER OF PHARMACOLOGY. "You are the best person in my knowledge to own this flag," Chang had written. "I hope, with this flag, I can be always at your side and can once in a while steal some idea to solve my [scientific] problems."

At the base of the bookcase Anne keeps the scrapbooks she has maintained over the years, filled with articles, press clippings, and photographs, the glue by now sometimes having lost its hold, leaving behind loose pictures and rectangular patches of crusty brown.

One photo, taken out of doors, shows Brodie and Udenfriend in the early days at NIH, around 1952, both wearing the pleated slacks fashionable at the time. Udenfriend, in a bow tie, stands with his hands resting on his hips. Brodie, dark hair slicked back, tie flapping in the breeze, squints under the harsh overhead sun. They are in their late thirties and early forties here, at the top of their form, shirt sleeves rolled up, expectant, brimming over with vitality.

Three decades later, a snapshot of Brodie would still show a vigorous-looking man. He is ruddy faced, with a full head of white hair, his eyebrows jet black, his chin still firm, his skin relatively smooth. His mind is as sharp as ever. He listens with ferocious intensity. And when he breaks into a big, toothy smile you can see a flash of the Brodie charm that has captivated so many.

But the strokes, heart attacks, and a case of Parkinson's disease have left him weakened. Even back in 1979, when he came up from Florida to the Goldwater reunion, Tom Kennedy remembers him as frail. Today, he walks haltingly, sometimes leaning on someone for support. He can read only with difficulty, word by painful word. He has trouble speaking, one of his strokes affecting the speech center of his brain, so that he stammers, sometimes catching on a word and being unable to complete his thought, his face contorting with frustration, or else substituting a simple word for a many-syllabled one.

He can no longer keep up with the scientific literature. "It's frustrating to have watched it happen," says Diane Russell of Brodie's physical decline. "He's become more of a recluse," not wanting to show his failing powers. Says Anne Brodie, "He's like an Olympic athlete who's had his leg amputated. He accepts it. But if he were capable of it, he'd be back in harness in a minute."

These days, Brodie sits for hours at a stretch in a little room at the back of the house, just off a walled desert garden.

To keep busy, he collects stamps and watches cable television. He still has trouble sleeping at night, so the TV is hooked up to an earphone, allowing him to listen, without fear of disturbing Anne, late into the night.

Sometimes, he pulls down the two thick books, bound in rich black leather, that he keeps behind the desk in his office, and sits looking at them for a long time.

14.
Epilogue: 1993

THE LETTER, on White House stationery and bearing the signature of Ronald Reagan, greeted scientists attending the National Academy of Sciences symposium and thanked them, in the nation's name, for their work. "You have something else in common, as well," it went on, "a legacy of scientific inquiry and inspiration from your honoree. . . . 'Steve' Brodie has spent a lifetime sharing his zestful energy, keen understanding of pharmacological chemistry, and penchant for relentless logic with generation after generation of budding researchers. . . ."

President Reagan's seemingly intimate knowledge of the personality and career of Steve Brodie, to be sure, may have been enriched by the kindly attentions of Mimo Costa and his friends at the National Academy. Costa directed a research institute established by the Fidea Research Foundation, and Fidea had invited scientists to Washington for "Neurochemical Pharmacology 1988: A Tribute to B. B. Brodie." There, under a tent erected outside the academy building, surrounded by Sid Udenfriend, Park Shore, Julius Axelrod, and dozens of other old friends, sat Brodie. He was eighty-one now, and looked it. It was April 29, 1988.

Two years before, Brodie, feeling isolated in Arizona, had moved with Anne to Charlottesville, Virginia, buying an apartment near the historic old center of town. Charlottesville is prime Civil War country, and even desultory browsing through the old town left him hooked, an instant Civil War buff. Costa, with whom his relationship had, if anything, deepened since they'd met at a Miami hotel desk almost thirty years before, began taking him to the bloodied battlefields of Antietam and Gettysburg, to the house in which General Lee had died, to all the holy Civil War shrines.

Whenever Costa came to Charlottesville for an outing, he'd find Brodie waiting, his hat on, cane by his side, ready to go. But this time, early in 1989, Brodie was sick in bed, so they spent the time watching college basketball. Later, getting up to go, Costa issued Brodie a stern warning: next time he'd better be up and about. But soon after, on a snowy day in February, he picked up the phone to hear Anne tell him she'd found Steve sprawled across the bathroom floor, dead of a heart attack.

He was cremated, his ashes spread across the ocean. His will made provision for the support of a B. B. Brodie Department of Neuroscience at the University of Cagliari in Sardinia, in Mimo Costa's home town.

•

Though ninety years old and stooped, James Shannon still looked dapper, almost preppy, in a sport jacket, blue oxford shirt, and perfectly knotted tie. He had recently moved from Oregon to a luxurious highrise retirement home in Chevy Chase, Maryland, looking out over woods and stream and stocked with card rooms, easel-equipped art studio, and wood paneled library. His hearing was almost gone. And he had occasional trouble remembering names—even briefly, those of Congressman Hill and Senator Fogarty, with whom he had happily schemed for the benefit of NIH during the 1950s and 1960s. But he kept busy, working on a history of medicine covering the years leading up to World War II.

His new home in Chevy Chase was only a few miles from Bethesda, where he had ruled NIH for fifteen years and where the NIH headquarters building today bears his name. No, he said, it was no big deal that they named it after him. Of course, if they'd named it after somebody else, he added, the smile spreading across his face its own short course in Irish charm, that would be a different story. . . .

•

In the summer of 1988, invitations, designed with slashing streaks of black, red, and silver, dissonant typefaces, and a number 8 metamorphosing into a drip of infinity symbols, beckoned friends and business associates to an August 8, 1988, party at the new laboratories of Peptide Design in Germantown, Maryland. It was, as Candace Pert recalled, "the booming biotech eighties." She and Michael Ruff, whom she had married in 1986, had left NIH and were now in business, reaping the benefits of investor interest in Peptide T, their stab at a treatment for AIDS.

By late 1985, the CD4 receptor in white blood cells had been identified as the AIDS virus's entry point into the body's immune system: A protein on the outside of the AIDS virus, known as gp 120, bound to CD4 and thereby began the cruel sequence of events that so often led to illness and death. Maybe, thought Pert (who with her colleagues had also found CD4 receptors in the brain), you could block AIDS virus binding by introducing some harmless substance structurally related to gp 120 and able to compete with it for receptor sites.

Pert and NIH colleagues ordered a computer to systematically compose the known structure of gp 120 with those of five or six dozen peptides—short, straight protein segments—known to play a role in the nervous system. Wanted: a best match. The computer spit out a structure four of whose eight building blocks, as it happened, were the amino acid threonine, which scientists represent with a T; Pert dubbed it Peptide T. She ordered it, and three kindred versions of it, synthesized.

In subsequent studies, three of these four peptides interfered with CD4 binding. And in another study at the Frederick Cancer Research Facility in Frederick, Maryland, the same three—Peptide T and two of its close cousins—blocked viral infection of human T cells. "The consistency was unbelievable," says Pert. "I thought that four weeks after that we'd be in *People*."

The early results, bearing the names of Pert, Ruff, and six others and published in the *Proceedings of the National Academy of Sciences* in December 1986, were enough to launch Pert on her quest for what she is not in the least reluctant to call an AIDS cure and to sustain her in the face of discouragement from the rest of the scientific community.

Yes, discouragement, say her critics—but only because other labs haven't been able to replicate her results.

Pert, though, insists that more central to Peptide T's lack of acceptance is that it ran afoul of an AIDS establishment already wedded to the palliative treatment AZT. This, plus the lingering effects of the Lasker flap on her personal standing among scientists, she says, led to Peptide T's premature dismissal. Major AIDS researchers, including what she makes sound like a veritable cabal at Harvard, have blocked research into Peptide T, even resorted to boorish and unprofessional behavior at scientific meetings, turning opinion against her and her discovery. As a result, she claims, one company that first backed Peptide T withdrew its support. And Peptide Design, launched with such fanfare to capitalize on Peptide T's promise, closed down. For a while, she and Ruff were reduced to working out of the basement of their home.

Pert gives this account with all her usual flair. Ruff, a bright, amiable immunologist a few years her junior, is quieter, if no less bitter. But he allows that their story does bear a faintly conspiratorial flavor, one that invites a natural skepticism.

Representatives of the AIDS-suffering community, eager to advance progress on any possible cure, however slim its prospects, have rallied behind Peptide T. "We believe that opponents of Peptide T's development," wrote a coalition of AIDS activists to Anthony Fauci, a high NIH official, "were motivated by prior interests in competing therapies, and that these conflicts of interest have distorted the peer review process" and its objectivity.

In 1993, Pert and Ruff were back in business with another company, Peptide Research, occupying a small suite of offices in a sweeping brick curve of a building along Parklawn Drive in Rockville, Maryland. A second company, Advanced Peptide, fueled with money from an investor Pert preferred not to name, was also sponsoring Peptide T study. And it had assumed the care of fourteen AIDS patients involved in an early study of Peptide T's safety and still alive four years later. Meanwhile, Phase II clinical trials, a Food and Drug Administration category of tests designed to test not just the safety but the clinical worth of a new drug, were under way at five sites. Results, said Pert, were expected by early 1994.

Asked about Peptide T's prospects, Sol Snyder–supremely guarded in discussing Pert, careful to disclaim expertise in the area of AIDS, and at first unwilling to offer any but the most constrained and neutral view—at last offered an opinion: He would be, uh, extremely surprised if Peptide T bore fruit. Many had tried to replicate her results; all failed. No conspiracy was out to undermine Candace, he said, his voice flat, his face a studied blank. The answer was much simpler: there was nothing, so far as he could see, to her claims.

•

When this book first came out, some of Sol Snyder's students had T-shirts printed up which read, in black lettering on an orange field, APPRENTICE. Snyder got a T-shirt, too; his read, GENIUS.

In 1993, his long face was more deeply etched, his hair was graying, and half-frame reading glasses rested on the bridge of his nose. But otherwise, Snyder at fifty-four, wedded to his thirty laps in the pool and five minutes in the jacuzzi each morning, looked remarkably trim and fit. To listen to him, little had changed over the years; he was still at Johns Hopkins, still presided over what he liked to picture as a modestly scaled research operation, still attended lab meetings, prowled the lab, talked science with his students.

But Snyder's quiet, comfortable office, the inner chamber of a suite of eighth floor offices bright with beige textured wallpaper and modern art, served more than ever as a nerve center of the neurosciences. Sol Snyder was an important, powerful, and public figure. His former students—like Anne Young, recently named chief of neurology at Massachusetts General Hospital, and Joseph Coyle, named to head a newly established Department of Psychiatry at Harvard Medical School—occupied key positions. But Diane Russell died in 1989, at the age of fifty-four from cancer. His book, *Brainstorming*, giving his side of the opiate receptor discovery, got major review attention. *Scientific American* and the *New York Times* each devoted scarce profile space to him.

Snyder's roots in classical guitar, his little mannerisms of face and hands, and his Julius Axelrod–inspired views of research creativity invariably figured in popular articles about him. But there was a simpler, more substantive basis, beyond the human interest fluff Snyder dismissed with a wave, for the attention accorded him: His lab churned out important discoveries, one after the other, with awesome frequency.

The biggest news was probably, well . . . the gases. Not gases, mind you, with eight-syllable chemical names newly discovered or synthesized, but rather gases prosaic and well-established enough to put an undergraduate to sleep—but which, through the work of Snyder and others, now loomed as key chemical messengers promising to open up a new world of neurochemistry.

The gas nitric oxide—*not* nitrous oxide, the "laughing gas" used in dentist's offices—is about as simple a chemical compound as you can imagine; a single nitrogen atom linked to a single atom of oxygen. An encyclopedia account written a few years ago would describe a boring, colorless gas first prepared in 1620, toxic in high concentrations, and notable for almost nothing else. Certainly no grander biological role could reasonably be foretold for a chemical so small and insignificant, not when most biological mechanisms are controlled by large proteins and other more complex compounds.

Or so it was thought until 1987, when a string of discoveries showed that a substance known to diffuse from blood vessels, causing surrounding muscle to relax and the blood vessels themselves to dilate—and until then known only as endothelial-derived relaxing factor, or EDRF—was, impossibly, none other than nitric oxide. Nitric oxide, it turned out, was the principal regulator of blood pressure.

Snyder leaped on this finding. If nitric oxide, this new chemical messenger, played so important a role in vascular function, he reasoned, might it not play a similar role in synaptic function, in how nerves communicated? Could it, for example, be in the brain? A train of experiments followed, and with them discoveries, and with them, scientific papers—adding to the more than six hundred already on his résumé.

"Nitric Oxide: First in a New Class of Neurotransmitters?" was the title of a review article Snyder wrote for *Science* in 1992. All the evidence, much of it coming from Snyder's lab, pointed to this simple gas as a key player in nervous system function, synthesized on demand within the cell. It differed markedly from neurotransmitters like norepinephrine, which Julius Axelrod had studied forty years before; but neurotransmitter it surely was. And soon Snyder's lab had revealed a similar role for another unlikely candidate, carbon monoxide, the odorless, poisonous gas conspicuous in automotive exhaust.

"Oh, Sol," says Julius Axelrod, recounting his protege's work with the gases, "he's fantastic."

•

In a photograph taken at an eightieth birthday party for him on September 18, 1992, and attended by former students, Julius Axelrod's son Fred sits to his left, Sol Snyder to his right.

Glowinski is there, and Hans Thoenen, and Wurtman, and the others, all with big smiles. . . .

A fine way, may we not assume, to conclude a distinguished career in science? The wizened, white-haired researcher, already a decade past retirement but still formally an NIH "guest worker," enjoys the plaudits of his devoted students one final time. Then, at last, he settles into a real retirement, a well-earned rest, and the free time to enjoy memories of a life in science well and proudly spent.

But this is a fiction, a scenario that never happened.

On a sunny spring afternoon in May 1993, Julius Axelrod heads off to the cafeteria. Joining him for lunch is biochemist Chris Felder. Back in 1985, Felder had called to invite him to give a talk at Georgetown, where he was then a doctoral student. He expected to face "three levels of secretaries" before reaching the Nobelist; Julie picked up the phone himself. Felder, thirty-nine, has been working with Axelrod, eighty, for six years now. He chats with him in the morning and over lunch in the cafeteria, finding him a bottomless source of fruitful scientific suggestions that always, it seems, turn out right.

Says Julie: "I'm still puttering along."

In fact, nine months after his birthday party—a trip to Europe later, a heart attack later, triple bypass surgery later, then another trip to Europe—Julie and his lab are in the thick of breaking science.

In 1992, a postdoctoral student named William Devane, pursuing work he had begun as a grad student at St. Louis University, teamed with a group of Israeli researchers to report the identity and structure of a substance found in the brain that binds to the same receptor as does marijuana; this was pot's equivalent of the endogenous ligand Hughes and Kosterlitz had found for the opioids twenty years before. "The cannabinoid ligand," *Science* quoted Sol Snyder as saying, using the scientific name for marijuana's active ingredient, was "an enormous breakthrough." It could speed the search for drugs with the known therapeutic effects of marijuana—painkilling, nausea-depressing, blood pressure-reducing—but without its high. *Anandamide*, they called it, invoking the Sanskrit word for bliss.

But what was the *normal* function of anandamide? Balance? Probably. Memory? Perhaps. What enzymes played a role in its production by the body. How was it metabolized? The cannabinoid gene had been cloned in Axelrod's lab. And anandamide was a derivative of arachidonic acid, long of interest to Julius Axelrod. Could he, Devane wrote him, join his lab? He could, and did.

In January 1993, soon after Devane's arrival, while in Europe for a scientific meeting, Julie felt pressure in his chest and found it hard to breathe. "It wasn't that bad," he recalls, "so I forgot about it."

But a few days later, back in the States, he came down from his apartment to his car and felt the pain again, this time more acutely. Someone called an ambulance. But tests on the scene revealed nothing. Take a cab to the hospital, he was told. So he did. The local hospital sent him to Georgetown Medical Center, where the cardiologist found his coronary arteries blocked, their capacity down to 15 percent of normal. He needed a bypass, where blood vessels from the leg or other parts of the body are snipped out and placed in the service of the heart.

Should he do it?

He called his son Paul. Paul called Sol Snyder. Snyder advised Julie to go ahead with the operation.

He woke, confused, with tubes sticking out of him. For days he was uncomfortably constipated. But after eight days in the hospital, he returned to his old two-bedroom apartment on Grosvenor Place, his home since before Stockholm. His wife, Sally, had died the year before. Now, his son Fred took care of him. Then, Fred's wife took over. Then Paul. . . .

That was February.

Now, in May, in the lab again, he unbuttons his open-necked short-sleeved shirt, pulls up his V-necked T-shirt, and shows the neat pink scar running down the middle of his white chest. "I feel fine," he says, smiling. And he looks it. He's a little thinner in the face and a little smaller in the gut than a few years before. And the leg from which they took the blood vessel for the bypass has atrophied slightly. But, altogether, he seems healthy and vigorous, just back from a European trip—to Budapest, Lausanne, Paris—where he'd given scientific talks. "What am I going to do?" he replies when asked about his rush back from surgery, "Sit around?"

A visitor finds him at his usual spot, seated in the tiny alcove just off the open doorway to the lab, chin sunk down almost to his chest, a scientific paper drawn up close to his one good eye. On a table just behind him are a coffee pot and tins of gourmet cookies a lab worker brings in to share. Outside, amid sinks, hoods, gradient cylinders, and racks full of test tubes, young people in jeans sit at the bench, measuring, pipetting.

As Julie starts to tell about his lab's most recent work and about Snyder's work on the gases, the phone rings. It's Sol.

On Julie's end, mostly silence, then an occasional soft, "Uh huh, uh huh." No gushy greetings. A routine call.

"How's the carbon monoxide work coming?" asks Julie.

More quiet. Listening.

"Yeah, you need the right person. It's not an easy problem."

Listening.

"So they accepted the patent, terrific . . . "

Sol, the son, tells of his work. Julie, the father, asks, encourages, listens.

Acknowledgments

MY THANKS GO, first of all, to Steve Brodie, Julie Axelrod, Sol Snyder, and Candace Pert. They let me into their lives, endured my questions, gave generously of their time, knowledge, and memories, and made available to me some of their personal papers and effects. My gratitude to them exceeds anything I can express.

Thanks, also, to the dozens of other scientists who recounted to me their personal and professional relationships with Brodie, Axelrod, and the others—and about whom many stories of scientific discovery remain to be told.

Thanks to Bob Goss and Sandy Jones, each for a crucial push when *Apprentice to Genius* was still just an idea; to Linda Nelson, for supplying boundless confidence, enthusiasm, and encouragement when mine flagged; and to Charlotte Sheedy for her excellent advice; to Cathy Mercaldi for rounding out my understanding of a key lab technique; to Babette Bilek for her invaluable library research; to Anne Brodie for her gracious hospitality in Tucson; to Barry Lippman, David Wolff, and Paul Heacock, for editorial criticism that, with humbling frequency, hit right on target.

Thanks, especially, to Elise Hancock into whose office at *Johns Hopkins Magazine* I tramped one day about three-quarters of the way through the writing. "Elise," I said, "here I've been writing about mentor relationships, and it just dawned on me that you're *my* mentor."

"I was wondering when you would realize that," she replied.

To my late grandfather, Saul Wolshine, whose work with wood and metal still influences mine with words. And to my parents: all through the researching and writing of this book, I felt so much of what they taught me as a palpable, everyday presence.

Thanks, finally, to Judy. She helped keep me steady. She endured me when I slipped. She closely, carefully read the manuscript as it came off the machine, always somehow responding with the precise balance of criticism, encouragement, and love that I needed.

And to David Saul Kanigel, whose own birth and first year of life hauntingly paralleled that of the book, and who helped make the writing of it a special joy.

Index